T0320418

Productive Safety Management

This book discusses the realm of operational risk management, exploring the intricacies of managing safety, production and quality simultaneously. It offers a fresh perspective on the dynamic and complex nature of risk, highlighting the ever-changing landscape that organisations must navigate. The reliance on current understandings of residual risk is deficient, particularly as systems of production are prone to degradation over time. This degradation leads to an increase in 'entropic risk', resulting in losses in daily production that, if left unchecked, could culminate in catastrophic consequences.

Productive Safety Management, second edition utilises practical experience to offer context and application to the concepts surrounding risk that are introduced. It explores the residual and entropic risks present in production systems before shifting focus to the same risks within organisational elements such as leadership, competencies, management systems and resilience. The degradation of these factors can lead to a toxic enterprise culture. Traditional risk management methods have resulted in the creation of functional silos. This book advocates for a multidisciplinary approach, positioning it as essential reading for the Fourth Industrial Revolution. In this era, the ability to effectively manage risks and capitalise on opportunities will be crucial for operational success.

This comprehensive title is designed for operational managers and supervisors, and risk-related professionals in engineering, OSH, environment and quality management.

Tania Van der Stap spent the last 20 years since writing the first edition of *Productive Safety Management* in managerial and technical positions responsible for safety, health and environmental management. Having experience in staff and contractor roles means she understands how to achieve results, whether within the organisation, owners' team, project team or as an external technical expert. The industries and organisations she's worked in have been diverse – gas, mining, exploration, construction, rail transport, engineering, agribusiness, professional organisations and regulatory authorities. She has in-depth knowledge of different strategies according to each organisation's level of maturity, leadership capability, resource availability and most importantly, the operational reality of the enterprise. Tania's qualifications are in commerce, which has throughout her career resulted in a business lens on operational performance. She is an unequivocal advocate of a risk- and opportunity-based approach to HSE, production and quality.

Productive Safety Management
How to Mitigate Residual and Entropic Risks

Second Edition

Tania Van der Stap

CRC Press is an imprint of the
Taylor & Francis Group, an **informa** business

Designed cover image: image credited to iQoncept; Shuttershock ID: 2273047259

Second edition published 2025
by CRC Press
2385 NW Executive Center Drive, Suite 320, Boca Raton FL 33431

and by CRC Press
4 Park Square, Milton Park, Abingdon, Oxon, OX14 4RN

CRC Press is an imprint of Taylor & Francis Group, LLC

© 2025 Tania Van der Stap

First edition published by Routledge 2003

ISBN: 9781032701578 (hbk)
ISBN: 9781032690247 (pbk)
ISBN: 9781032701561 (ebk)

DOI: 10.1201/9781032701561

Typeset in Times
by codeMantra

I'd like to dedicate this edition to the people in industry operations whom I've worked with in the last 20+ years since writing the first edition of Productive Safety Management.

Both support and criticism have motivated my professional journey. I'm particularly grateful to those who trusted me by sharing their successes, challenges and frustrations with managing safety and operational risks. This edition is a culmination of what I've learned about achieving a safe, productive, quality job in numerous industries.

My hope is that readers will make a transformation from 'silo- based thinking' to 'risk-based thinking' at work and as a life skill.

Contents

Figures

Tables

Foreword

As an engineer in the occupational safety and health profession for several multinational energy businesses and later business owner employing people in the manufacturing of ultrasonic equipment and in agriculture operations, my 40 plus years of education, professional development and experiences impressed upon me the awesome accountability for managing risk safely as a pre-requisite to achieving production output goals and quality work. The ever-present business decision and execution requirements demanded an awareness and balance for all stakeholders' perspectives, their motivations towards risk acceptance and rewards. So it is my privilege to introduce this second edition of *Productive Safety Management: How to Mitigate Residual and Entropic Risks* as a must read.

This book represents the author's decades of cumulative education and work experiences, focusing on practical application as needed by business owners and operational leaders as opposed to purely academic theory and teachings, gleaned from her perspective working with a myriad of business owners, operations managers and risk-related professionals. She takes further steps forward from the first edition of *Productive Safety Management*. In this second edition, one can find a voice that's operational, not academic, and much more than a slogan, 'Safety First'. The author seeks a transformation from 'safety- based thinking' to 'risk-based thinking' at work and as a life skill for owners, engineers, front-line supervisors, employees, contractors and others having a stake in the enterprise, in their communities in which they work and for their families.

The book and its contents are organised and succinctly presented in a manner well-suited to those that can make a difference in terms of risk and reward while cognisant of the legislative and bureaucratic regimes that can be viewed as barriers in businesses today. Contained in the chapters, one can sense the call to action for reassessment of commonly accepted views of risk assessment and mitigation and how processes, technology, the physical work environment and people all come together and sometimes collide in their intents, to sustain a successful operation. Here in this second edition are presentations on topics with discussions, challenges and potentially, revelations.

The author brings forward for consideration of the Entropy Model and offers strategies for management. Not to be overlooked is a spot-on need to understand the impact of organisational maturity and the will and ability to mitigate risk. In other chapters, the discussions turn to processes and technologies, both never static, but always in a dynamic state of change, and thus presenting ever-shifting probability and frequency of risk and acceptance. Coupled with the physical environment impacting operations, we see risks, sometimes called 'Acts of God', that can be more challenging and can enable excuses by owners and leaders for failure to act on risk.

The author provides a view to risk quantification and management strategies dispelling assumptions, providing case studies and offering residual risk management (RRMS) and entropic risk prevention strategies (ERPS). Even the most successful businesses are subject to continual and accompanying risk in the ever-changing dynamics of leadership and organisation capacity for risk decision management. As businesses plan for and sustain a high-reliability organisation (HRO), existing and new risks continually "creep" into systems, processes, equipment, the work environment and people, the development and implementation of a fit-for-purpose business model addressing residual and entropic risks become ever more urgent.

The early chapters focus on the residual and entropic risks associated with systems of production, whilst the latter chapters propose that organisational factors such as leadership, competencies, management systems and resilience also carry these risks. Collectively, these determine an organisation's capacity to manage risks and exploit opportunities. Of great concern is the tendency towards cultural degradation which has been the case prior to many catastrophes. The book covers more than disaster prevention and talks pragmatically about operational failures. On a day-to-day basis, operations can experience losses such as safety incidents, production downtime and quality deficiencies, because of poorly managed risks, so the book provides strategies for timely intervention and business improvement.

The author is to be commended for writing this second edition. This book is more than theory and management; rather, it is a practical instruction to assess actions in recognition of ever-changing risks in business and operations. This is a book for anyone genuinely interested in the protection of our greatest asset, human life, whilst concurrently achieving production output and quality work. For the above-stated aspiration and intent, this book is a must read and application.

Michael W. Thompson, FASSP, CSP (ret.)
Freeport, Illinois, USA

About the Author

Tania Van der Stap has spent the last 20 plus years since writing the first edition of *Productive Safety Management* in various management and technical roles covering the disciplines of occupational safety and health (OSH) and environmental management. Her positions have included both staff and contractor engagement models, giving her an insight into being part of the organisation, owners' team, project team and external technical expert.

Likewise, the industries she's worked in have been diverse, including oil and gas, mining, exploration, project development, construction, rail transport, engineering, agribusiness, professional organisations and regulators. This exposure has resulted in a deep understanding of the need for different strategies according to each organisation's level of maturity on the risk management journey, leadership capability, resource availability and, most importantly, the operational reality of the enterprise.

In 2023, Tania was engaged by the Institution of Occupational Safety and Health (IOSH, UK) to write textbooks for their degree-level course in OSH Leadership and Management. This work, like this edition of *Productive Safety Management*, is a culmination of extensive experience as well as contributions to thought leadership since the first edition. Tania has written several journal articles as referenced throughout and "Chapter 29: Risk Leadership: A Multidisciplinary Approach" for the American Society of Safety Professionals book *Safety Leadership and Professional Development*.

Tania started her consulting business, Align Risk Management based in Western Australia, in 2003, following the global publication of the first edition. She wrote the book considering several fatalities in the mining industry in the 1990s and concerns about the impact of production pressure on workers' safety. These concerns remain, balanced with contemporary challenges of the over-bureaucratisation of OSH and operational risk management systems. Tania has extensive practical experience in the development of right-sized, risk-based OSH and operational management systems, in collaboration with risk owners with a focus on 'needs' not 'wants'.

Perhaps ironically, Tania's university qualifications are in Commerce not OSH, which has throughout her career provided a business lens on the organisation and been the basis of how she operates in the work environment. For these reasons, she is an unequivocal advocate of a risk- and opportunity-based approach.

Introduction

"An idealist is essentially risk-loving with respect to the achievement of the ideal" (Cambridge Dictionary, 2024a). The hook of the quote is 'risk-loving' which is a counter-intuitive opening for a book that on face value may appear to be just another publication about workplace health and safety. The author is unashamedly an advocate of embracing risk (and opportunity) as a realistic and essential element of human industrious endeavour. This has unfortunately been lost by the occupational safety and health (OSH) profession in which the author has worked for over 20 years since writing the first edition of *Productive Safety Management (PSM1)*.

In workplaces, there are too many conversations about what can't be done for safety reasons as opposed to what can be achieved – safely, productively and with a quality result worth celebrating. The good work done by many OSH professionals has become overshadowed by a reputation as 'nay sayers', 'wet blankets' and obstructors of production. From experience, this has resulted in having to earn and sustain credibility on every new contract, of which there have been many across various industries during a 20-year career.

Credibility is a high value commodity, especially in hazardous industries.

Without it, a working person may be perceived as inept and vulnerable. At the professional level, reasonable success is dependent on being a realist – *"a person who tends to accept and deal with people and situations as they are"* (Cambridge Dictionary, 2024b). Great success needs more. It needs belief and vision. This book is a blend of realism with idealism applied to the management of risk, particularly in operations, for good reason.

The overarching objective of this book is to pursue health, safety and environmental (HSE) performance concurrently with production output and quality work.

'Production' is target-driven, for instance, X number of widgets off the processing line or Y tons of ore from a mining operation. Throughout this book, it has equivalents across different industries. For instance, in the public sector, production equated to 'productivity'. In construction, it is likened to 'schedule'. In the food and beverage industry, it's customer service and satisfaction. In health services, it relates to patient care. Fundamentally, 'production' is core business.

Realism is essential to understand risks associated with production, whether you're an engineer, operations manager, supervisor or from support services such as OSH, environment, quality, procurement, contractor management and so on. In *PSM1*, it was explained that society's current approach to risk management is deficient due to the failure to recognise and address the tendency of systems of production to degrade

(Mol, 2003). This flawed perspective still prevails. The topic will be explored in depth in *PSM2* together with pragmatic solutions to improve capacity to manage risk and workplace practices Operations have the most tangible risk profile. How well this is managed greatly influences the performance of the enterprise and in particular, its ability to mitigate losses.

A shift from 'safety-based' thinking to 'risk-based' thinking is needed to optimise performance and pursue potentially conflicting organisational objectives concurrently, such as production targets and incident prevention. Idealism is also vital to embrace and enable human potential – to adopt a mindset of what can be achieved and to provide a working environment for capacity building. The journey described in *PSM1* from 'safety' to 'risk' as the driver of a multidisciplinary enterprise strategy is mapped in *PSM2* with contemporary application and the benefit of a wealth of operational experience.

Stepping away for a moment from the workplace and looking more broadly at community, the essence of human existence that captures our imagination is the need to survive and the desire to thrive. These themes permeate history and continue to underpin society's storytelling. We're attracted to and in fact rejoice when events of intense human adversity are overcome such as when the 12 boys from the Wild Boars football team were rescued from Tham Luang Cave in Thailand after being trapped by heavy rains and floods in June 2018 (BBC News, 2018).

Our appreciation of human endeavour is warmed by great achievements. We wondered years ago when world records at great sporting events would no longer be attained and yet, previous best times continue to be broken. Our everyday lives are being transforming by technological innovation and advances. The need to survive and the desire to thrive are enduring qualities, and of course, we each have many examples to draw on personally and professionally.

In today's organisations, these fundamentals are framed to some extent by 'sustainability' and 'resilience'. From a humanistic perspective however, there's a shortfall in any semblance of collective rejoicing in many workplaces. The question should be asked, "Is something missing?" Does an experience at work have to be grandiose to be worth celebrating? From experience, this is an area of management that's lacking. It can't be simply glossed over with an end of project dinner or a team golf day. More positive experiences should be happening day-to-day for managers and workers which might be described as "Felt Success". Modern workplaces seem to be desolate of heartening human endeavour and the OSH profession has some accountability for this. The growth of 'safety' as an industry has a downside that has become the modus operandi of the profession.

Bureaucratisation of safety and the culture of risk aversion is a disservice to the workforce (managers and workers alike) and to progress.

In recent years, the philosophical movement away from complex OSH management systems and processes has gained traction amongst safety professionals, led by eminent academics in the field. This has focused attention on "Work as Imagined" (WAI) as opposed to "Work as Done" (WAD) thanks to Sidney Dekker and others. WAI represents the development of OSH processes by those who aren't directly involved in the work but may be considered subject matter experts in risk management such as OSH professionals. WAD is the operational reality in which OSH and other risks must be managed to prevent injuries and incidents. It's generally agreed that there's a gap that can render OSH paperwork inefficient and ineffective. The distinction between WAI and WAD is not new as evident in *Barcock v Brighton Corporation* [1949] KB 339. Hilbery J noted:

> a system of work is not devised by telling a man to read the regulations and not to break them
>
> It is no use for a master to say in court 'I discharged my common law duty because I put down on paper a safe system and put it into the hands of the man if for years they showed him by the action of those who were his superiors that the work was not done in that way, need not be done in that way; and could be done in another way which involved danger.

(Lawinsider.com, 2023)

This is a very powerful quote in the context of OSH management. With the gap between WAI and WAD now accepted as problematic, a new dilemma has emerged for organisations who've invested heavily and embedded their OSH and operational management systems. How can they firstly, close the gap and secondly, remove processes that are of low value in terms of risk mitigation? As explained in this book, the over-reliance on additional administrative controls as corrective actions following incidents may have created greater complexity for the supervisors and workers who must implement and comply with these rules and procedures. It's little wonder that "Felt Success" is so elusive.

The reduction of onerous processes presents a very real challenge for corporate and operational managers. How can they justify the removal of existing safety processes or layers of risk assessment and/or controls, to achieve simplification having been the instigators of such bureaucracy? This is in the context of increasing corporate accountability and also, in some jurisdictions, exposure of individuals to risks of penalties and potential jail sentences from OSH legislative duties imposed on enterprise 'officers'. OSH management systems and practices have developed to both protect workers from harm and the organisation from statutory penalties and reputational damage. Both guard against these negative outcomes and perpetuate a culture of risk aversion.

Does this mean that organisations are stuck with a status quo of burdensome OSH systems? Improvements in productivity and minimisation of waste are paramount in today's economy in which positive environmental, social and governance (ESG) outcomes are expected of corporations. The question to the OSH profession is how will they adapt to deliver assurance based on prescribed legislative requirements as well as add to productivity improvements? The implication is the need for change which means:

The OSH Profession needs to loosen their grip on risk aversion and 'what-if' stories told to justify outdated, redundant approaches to OSH management that border on fearmongering.

Whilst there's general agreement of the need for transformation in the OSH profession, the missing piece is 'How'. The primary aim of this second edition is to explain how to take a risk- and opportunity-based approach to 'The Work' to pursue HSE performance, continuity of production and quality outcomes concurrently. That aim requires a realist's approach to risk management. Essentially, this includes tackling the following issues:

- Challenging cultural rhetoric such as 'Safety First' and 'Zero Harm';
- Acknowledging that risk is dynamic not static so today's risk review is likely irrelevant tomorrow;
- Disciplinary silos and their respective approaches to risk management are creating blind spots that can lead to losses including catastrophic events;
- Adding lower order controls such as administrative procedures may be increasing complexity that raises rather than reduces operational risks in the flawed belief that the new risk level is 'as low as reasonably practicable' (ALARP); and
- Residual risk defined by 'ALARP' doesn't clearly consider the fact that systems of production tend to degrade.

PSM1 was written considering several fatalities in the mining community where the author resided. It was apparent that across the industry in the 1990s (and prior to then) that production pressure was a challenge to manage and often presented a hindrance to effective safety management. The problem was how to overcome the conflict between safety and production. *PSM1* proposed that production, safety and quality aren't ends in themselves but outcomes of effective risk management. Consequently, the sub-title of *PSM1* was, "*A strategic, multidisciplinary management system for hazardous industries that ties safety and production together*".

At the time, it wasn't common practice to integrate human resource management (HRM) strategies such as leadership, competency training and development with safety management. This was a significant gap attributable to the immaturity of both professions. For example, HRM was in many organisations only just emerging from personnel management which was primarily concerned with the employment process and payroll. OSH was also rudimentary and non-strategic, focused on tasks and fault-finding at the operational level. To overcome the gap, *PSM1* integrated OSH, HRM, quality, environmental management and engineering to provide a whole business approach to effective risk management.

There were two key models that underpinned the management system presented in the previous edition. The first, 'The Entropy Model', which is explained in detail in Chapter 1 of this edition, illustrates that there are two types of risk – residual risk and entropic risk – that can lead to losses in production, HSE and/or quality performance. It also provides a four-step risk management strategy for mitigating these risks that can be applied strategically and operationally.

The second model, the "Strategic Alignment Channel", presented again in Chapter 2, explains the need for the enterprise to achieve alignment across three areas. These are firstly, of the organisation to the external environment, secondly, internally of the enterprise's physical, financial and human capital, and thirdly, internally between organisational and workforce values. The original "Channel" in many ways pre-dated today's approach to ESG frameworks. It's the Entropy Model however that has emerged from initial publication in 2003, as having significant relevance to today's

OSH profession and other risk-related disciplines. As explained in Chapter 2, the organisation continues to require the input of multiple functions such as engineering, HRM, OSH, quality, environmental and operational management to achieve effective risk mitigation and opportunity exploitation across the enterprise and throughout its lifecycle phases.

The second primary aim of this edition is to establish a renewed ideal for the management of risk and opportunity in organisations that enables great outcomes for people by transforming some entrenched bad habits and instilling good ones. The first one of these is blame. Other authors including Dekker (books including 'Just Culture' and 'Stop Blaming') have covered this topic from a psychological and sociological perspective in relation to OSH practices that result in blaming the worker (unsafe acts) for accidents. In this book, the fallibilities and capabilities of human beings are discussed from an objective risk-based perspective. Human resources are one of four systems of production that entities employ to achieve outputs (products or services). This systems-driven approach reduces biases.

Shifting blame to management is also unconstructive. Managers and supervisors might want to promote employees' ideas and create an open, honest culture but may not be empowered to act on the input from their team or they may feel compelled to adopt a short-term outlook to work (Sherf et al., 2019). Organisational conditions such as these were explained by the Alignment Fallacy in *PSM1* and will be revisited in this edition. The second bad habit addressed in *PSM2* is under-valuing people but with a caution.

Catchcries such as 'Our people are our Greatest Asset' are as superficial as 'Safety First' when tough decisions must be made about resourcing levels and production targets.

Throughout this edition, where there's rhetoric in current industry language, it will be called out. This pragmatic approach coupled with an objective risk-based approach to HR risks is balanced with humanistic values and high regard for managers and workers alike. The author honours the many people worked with over the years but more importantly, it feels true to self. After all, most people are doing the right thing for the right reasons.

The rest of this introduction addresses how these aims will be delivered. Chapter 1 will provide a more detailed explanation of why the Entropy Model has increased in relevance plus the contents of more recent republications including the American Society of Safety Professionals' (ASSP) book, *Safety Leadership and Professional Development* (2018), 'Chapter 29 Risk Leadership: A Multidisciplinary Approach'. In 2020, the model was expanded to address organisational and leadership factors in the ASSP's Professional Safety Journal, *Safety and Entropy: A Leadership Issue* (Grieve and Van der Stap, 2020).

More broadly, there has been a shift towards a risk-based overview involving integration of disciplines. The overlaps of OSH and HRM are now recognised including leadership and culture as part of human capital development. The work of numerous subject matter experts (Klussmann, 2009; Kymal et al., 2015; Pardy and Andrews 2009) have acknowledged the benefits of integrated management systems (IMS).

These include cost and resource savings with improved financial performance. From a risk management perspective, there's better risk prioritisation, positive corporate reputation, improved communication and efficiencies in business operations into a single business management system.

The recent changes to International Standards have directed organisations to this risk-based approach as evident in ISO 31000 Risk Management, ISO 45001 OSH Management Systems, ISO 14001 Environmental Management and ISO 9001 Quality Management. The standards previously didn't prescribe the need for opportunity analysis and stakeholder needs considerations as part of enterprise strategy development. Many segments of industry are receptive to this concept of a fully integrated risk management methodology addressing systems risks however, they don't necessarily know how to best achieve this. This topic will be discussed in Chapter 2 and developed throughout subsequent chapters. Overall:

It's a necessity that businesses balance potentially conflicting objectives such as production, quality, OSH performance and cost to ensure short-term viability and longer-term sustainability.

In the context of conflicting goals, there are numerous qualitative and quantitative tools used to drive risk-based decision-making to seek optimal outcomes. The potential struggle between safety and production must be addressed to achieve safe, efficient, quality-driven work and decision-making from the board to the operational level. There remains, however, a gap between high-level risk management principles (societal and enterprise) and operational practice. This is most evident in high-risk industries where fatalities, serious injuries and debilitating industrial illnesses continue to have the greatest rates. The fundamental questions are: 'What is this concept of alignment?' 'Is alignment being achieved in organisations in relation to safety and production? 'If not, why not?'

The impact of this emerging risk-based approach is that for implementation to be effective, discipline silos within organisation need to be broken down. Hollnagel (2021) reinforced this and highlighted the negative impact on culture with the potential for conflict over safety versus quality versus production versus reliability cultures. The resulting confusion and intra-disciplinary non-alignment require a broader understanding of what connects failure and success in each and all disciplines. According to the principles of the Entropy Model:

The key issue that management should be worried about is evidence of losses. Production downtime, quality deficiencies and safety incidents are all losses connected by ineffective risk management.

Discipline-based silos can impede the enterprise from seeing the operational reality with the risk of being dominated by the loudest voice which is usually core business (production). Within disciplines, the focus can be too narrow. For instance, the OSH profession's primary work method is to identify 'hazards', assess the risk and advise on suitable controls, then to monitor the work to ensure that such controls are effective. The problem is that work is dynamic and therefore 'hazards' emerge then disappear and new 'hazards' cycle through. An example is the arrival of building materials on a construction site and the production of waste, both of which constrain the active footprint. This may then have effects on access and egress around the work area. Construction involves multiple tasks by different contractors and some of these activities being undertaken simultaneously (SIMOPS) may be incompatible, for instance, overhead crane operations adjacent to construction crews working at ground level. These dynamic, complex and ever-changing risk profiles are discussed in detail in this edition.

Updates to ISO standards, legislation in some jurisdictions and industry-driven practical planning, are struggling to shift the obsession with 'hazards' to broadened 'hazard and risk'. More needs to be done to expand the risk perspective.

The Work is dynamic not static. The focus needs to be on risk not simply hazards.

There may be known 'hazards' that need to be mitigated but it's the changing interfaces between human resources, processes, technologies, and the physical work environment that can fluctuate and result in injuries (from an OSH perspective) but also losses (from an operational perspective -production, quality, cost). These are all part of normal working situations and don't need to be extraordinary to result in a negative outcome.

The Entropy Model will be used in Chapter 3 to discuss processes (how the work is done) and these constantly moving parts that affect organisational performance across functional areas. An important element will be the impact of complexity. Adding further rules and procedures because of an incident investigation may increase rather than reduce the risks that contributed to the undesirable event. This will challenge current assumptions about 'as low as reasonably practicable' (ALARP). Are enterprises being realistic about their residual risks when they're relying heavily on procedural controls to manage these risks? Such controls depend on human performance which is known to be fallible.

Traditional risk assessment processes factor in exposure. This isn't simply duration. It's a sum of a complex set of variables associated with systems of production. This will be discussed in depth and used to question current incident investigation processes that place too much emphasis on human error. Case studies will be used to analyse the combinations of risk sources that can lead to losses and explain where human error fits within the total risk profile of the workplace.

Chapter 3 will touch on some key definitions within OSH legislation and explain why these present opportunities for businesses. The cases for minimal compliance and beyond compliance will be presented. The latter will introduce the concept of a Principal Risk and Opportunity Strategy proposing that it may enable enterprises to adopt the behaviours of high-reliability organisations (HROs) and thereby pursue improvements across multiple functions.

In many industries, OSH procedures have become increasingly complex, so a sense-making framework is needed to right-size management systems, documents and tools.

*Risk-based management systems should be designed to ensure safe,
efficient, quality-driven work which concurrently protects operations from
chaos and complexity.*

A case study from the utility industry will be used to evaluate four categories of
intervention and how these can be applied. These are 'simple', 'complicated', 'complex' and 'chaotic'. Traditional OSH risk management practices will be challenged in
terms of the application of ALARP. The 'ALARP Assumption' explains that operations can inherit deficiencies in design because of project cost pressures. A warning is provided against low-value thoroughness – adding further layers of low-order
controls with diminishing return on risk reduction, and the potential for complexity
to make risk rise as workers become overwhelmed by rules. This organisational risk
aversion may introduce risk transfer where increased thoroughness, which is time
consuming, results in unforeseen consequences such as production pressure and a
negative impact on the workplace culture.

In Chapters 4–6, the other systems of production residual and entropic risks will
be explored in detail. Chapter 4 covers those risks associated with technology, which
includes plant, equipment and tools (both hardware and software). The discussion
will go back to the basics of engineering decisions that affect safety in design, the
quality of components and standards of construction. The aim of ALARP should be
to minimise the residual risks inherited by operations and the potential entropic risks
(caused by degradation) that will develop through the technology's lifecycle.

It will be proposed that high-risk industries have an unacknowledged predisposition to failure because entropic risk prevention is not part of the language of current
organisations. As a solution to this gap, the five principles of HROs will be introduced with particular focus on their preoccupation with failure. Although there are
numerous disasters from which the industry should have learned and attained deeply
embedded and practised risk management capabilities, catastrophic losses continue
to occur. This points to the need for greater learning across industries supported by
global communication capacity to prevent disasters. Historic case studies will be
referenced to support this argument.

The chapter on technology wouldn't be complete without an exploration of the
rate of change in society. This calls for transformation on several fronts. The first is
professional adaptation.

*Risk-related professionals should be enlisted out of their narrow discipline
to a 360-degree perspective of their organisation.*

This is required to better understand the impact of these technological changes.
In this mix, in relation to both traditional and future systems of production, is the
human-technology interface and the interdependencies that cause risk to be dynamic
and complex. Secondly, some of the key issues in managing technological risk that
currently face operations remain highly relevant. These include supply chain relationships, standardisation where possible and both in-house and outsourced maintenance. The benefits and disadvantages of various strategies are considered from an
operational risk perspective.

Thirdly, the extent of the Fourth Industrial Revolution will be visualised with a particular emphasis on the impact this will have on operational risk management. Whilst the pervasiveness of this revolution is difficult to fathom, there are some key themes that have emerged. The input of a specialist researcher in artificial intelligence was sought to complete this section of Chapter 4. The trends that will be discussed include human-machine cooperation; the balance of power in human-machine systems; nonlinearity of complex machine adoption; trust between agents and reliability of the whole system; and privacy versus efficiency. The revolution is revisited in the subsequent chapters about the physical environment and human resources.

To close out Chapter 4, the efficacy of traditional enterprise risk management practices is contested. The solution proposed is a shift towards 'sense-making' as the foundation of risk-based decision-making. This provides a baseline for later chapters that explore choice selection at the leadership, supervisory and worker levels.

Chapter 5 covers the residual and entropic risks associated with the physical environment system of production. The key theme of HSE in design is followed up from the technological discussion in Chapter 4. The impact of engineering for HSE begins with minimising the residual risks inherited by operations and the potential entropic risks that will develop through the facility or site lifecycle.

Current organisations struggle to consider the wider risks to their business units that arise from the natural physical environment. There are significant constraints in risk assessment which can lead to 'black swans'. These are unknown unknowns and result from incomplete data, such as the 2011 Tohoku earthquake and tsunami that caused the Fukushima nuclear accident. In this mix, there is continuing confusion and the re-emergence of terminology acknowledging that some risks aren't within human control. If the worker or the manager can no longer be blamed, must God be the blame?

It's critical that the concept of 'Acts of God' in risk management practices including OSH is clarified.

When does a root cause of an undesirable event become an 'Act of God' compared to a system deficiency? These 'Acts of God' are sometimes used to explain catastrophes and extensive fatalities so, as explained in Chapter 5, the management of physical environment risks must include fatal risk controls. Monitoring regimes and emergency response capability are critical on the preventative and recovery sides of a bow-tie risk assessment, respectively. Operationally, these two mitigation strategies contribute to and demonstrate tangible organisational capacity.

The fundamentals of residual risk management will be explained in relation to the physical workplace. Poor initial site design is a major long-term contributor to sub-optimal productivity, safety, environment and quality performance and potential lost opportunities when modifications are made, for instance, when existing infrastructure creates a constrained footprint for expansion. In many organisations, there are avenues to leverage practices used by industry leaders. For instance, small to medium enterprises (SMEs) can apply improvement strategies such as the Lean Production Principle of 5S within the physical workplace.

Practical benefits of Lean Production are explained along with a cautionary word about being too pedantic about 'compliance' during day-to-day operations.

The concept of the physical workplace having a 'saturation point' where chaos transitions from supporting efficiency to becoming inefficient will be explained. In this context, the current approach to undertaking workplace safety inspections is challenged.

The truth is that 'organised chaos' is a natural characteristic of how work gets done. Operational personnel will often use this term to describe the reality of day-to-day work.

Chapter 5 will conclude with a visualisation of the physical workplace changes resulting from the Fourth Industrial Revolution and the impact these will have on operational risk management. This considers what constitutes the workplace currently and into the future. The characteristics of the gig economy are particularly interesting as is the rise in home-based work since COVID-19. Autonomous technologies are removing some workers from the hazardous environment. The final topic covered in this area is the adaptation of the workplace for an aging labour force.

The final chapter (Chapter 6) on systems of production will talk about human resources residual and entropic risks. The confusion which abounds in the OSH profession calls for a shift to a systems approach to HR risks and move beyond the human as hero or hazard. This will be contextualised along with the psychosociological lens currently being applied to OSH and operational risk management. It's important to step away from associating the technical process of assessing HR risks and confusing this with any form of discrimination. The objective is to protect people from harm and enable productive work, not to treat them unfairly or downplay moral and ethical principles.

Chapter 6 will introduce the influence of mathematics on human performance that supports the concept of rising entropy as a risk. Analysis is enabling more objective HR risk assessment. Applied mathematics including Shannon's Entropy Method results in better assessment of failure modes and effects. It looks at the degree of disorder in the system (of which a human may be a part) to identify unexpected trends that may present risks. Human fallibility is accepted as a fact and doesn't interfere with the evaluation methodology.

In *PSM1*, the highest risk group identified were young, inexperienced workers. This is still the case at the society level but the increase in incidents and injuries in older workers as evidenced by statistics suggests that this demographic requires specific risk management strategies. This becomes an important consideration in Chapter 7 in which a semi-quantitative risk assessment method is proposed to evaluate the residual and entropic risks of each of the systems of production including human resources (as described later in this introduction).

In Chapter 6, some of the traditional OSH-related human risks are discussed such as fatigue, mental ill-health and stress. A table of residual and entropic risks with applicable risk management strategies will be provided to avoid replicating the detailed discussion in *PSM1*. In this edition, the commentary turns to clarifying those risks that are an organisational accountability versus those that are the individuals' accountability. Chapter 6 also introduces the concept of leadership degradation.

A significant gap in current management approaches is the failure to monitor and to act effectively to support personnel in influential leadership positions through phases of degraded job performance, using a risk-based approach.

Recent changes to some OSH legislations have included psychosocial hazard management but these are insufficient from a holistic perspective. Degradation in the leadership ranks can have a ripple effect to peers and subordinates which should trigger corrective action by senior management and a 'maintenance' strategy to assist the affected person to improved job performance.

A further area that requires an objective risk-based approach is the clarification of what constitutes a 'safety violation'. This will be considered within a production-driven organisational culture and strategies will be suggested to ensure fair outcomes with optimal individual and enterprise learning. Violations are behaviours that are a clearly defined subset of HR risk that need to be evaluated within a systems approach. In practice, what constitutes a 'violation' can't simply be dismissed as intent. However, a warning is also given against unclear boundaries of accountability and responsibility. These may result in unexpected cultural consequences.

An individual or group in collaboration mustn't be able to blame the 'system' or 'production-pressure', as a means of avoiding accountability for a violation that is supported by objective evidence.

Chapter 6 which is the last of the series on systems of production risks will raise some of the key developments in personalised technologies emerging in the Fourth Industrial Revolution. The dialogue includes the use of exoskeletons for biomechanical support and the integration of wearables into OSH and operational management systems. The benefits and disadvantages will be presented along with social issues of privacy and misuse of data for individual performance monitoring.

Much of the contents of Chapter 7 will cover risk quantification and management strategy taken from *PSM1*; however, this will be refined with the benefit of 20 years of operational experience. From an academic perspective, the proposed methodology is yet to be tested empirically so the reader is encouraged to consider the logic which has been applied. The purpose of the approach is to promote healthy cynicism towards current risk assessment methods, particularly, reliance on risk matrices, and to provide an alternative. From practice as a risk workshop facilitator, participants' discussions can drift into debate and personal stories of incidents through their individual perception of what they consider to be acceptable versus unacceptable risk. The skilled facilitator must bring the group back to the purpose of ranking risks relative to each other. Most operational risk assessments using the matrix generate a risk register which is primarily 'work-as-imagined'. The semi-quantitative risk assessment method in Chapter 7 will break risk down into its component parts using real-work scenarios. This aims to provide a more pragmatic understanding of the complexities and interdependencies between systems of production than applying the risk matrix to 'what if' scenarios.

The deficiencies of risk matrices will be presented, and a case put forward for a mental model for assessing residual and entropic risks of the four systems of production, using eight risk scorers. These will factor in primary risk parameters and risk modifiers. For instance, for process residual risk, the energy level is proposed to be the primary risk parameter. The complexity of the process is the key determinant of the process entropic risk. The method that will be presented has been named the 'Dynamic Combinations Risk Method' (DCRM) to acknowledge that operational risks are dynamic and combinations of systems of production including the interfaces between these systems.

In this edition, the 'ALARP Assumption' will be added. This will be revisited in Chapter 7 to re-emphasise serious concerns about the organisational habit of simplistic corrective actions following incidents. Do more procedures and rules really reduce the risk or only make the work more difficult to achieve efficiently? Are there any subsequent negative consequences for the culture and operational buy-in?

The 'ALARP Assumption' suggests that as operations mature, there's a tendency after incidents to add more low-cost, lower order procedural controls that may actually increase the risk rather than reduce it.

In Chapter 7, DCRM will be applied to the case study of the BP Refinery Explosion and Fire that occurred in Texas in 2005. The residual and entropic risks that were present prior to the disaster will be identified using the regulator's report and analysed to consider how these contributed to the catastrophic event. Figures 7.2 and 7.4 will illustrate the scoring of the residual and entropic risks, respectively. The eight risk scorers (residual and entropic risks for each of the four systems of production) include risk modifiers which either increase or decrease the risk. For instance, simplification and standardisation are proposed to reduce process entropic risks, in other words, lowing potential for deviation from well-planned activities that might otherwise introduce degraded states.

It will be proposed that a risk assessment team would better understand the key strategies for residual risk management and entropic risk prevention if the primary risk parameters and risk modifiers became part of analysis discussions. These could then be used to identify more effective risk management strategies than the current practice of listing controls in the Risk Register then ascribing a residual risk score after controls are in place. It will be proposed that if DCRM is a valid risk assessment approach, then it would follow that the Total Risk Profile of an operation would comprise the sum of risks of the activities undertaken.

Chapter 8 will introduce the Entropy Model applied to the less tangible factors that determine capacity. These will be identified as capacity builders – leadership, competencies, management systems and resilience with resourcefulness. It will be explained that such factors have residual risks and can suffer from degradation that can lead to organisational losses. Many of the disasters that will be discussed throughout this edition are of organisations that had serious deficiencies in leadership and culture leading up to their catastrophic events. From experience, part of the problem lays in leadership expectations and how these are framed compared to how leadership is delivered in the workplace. Perhaps, there's such a thing as 'Leadership-as-Imaged' versus 'Leadership-as-Done'?

Some of the rhetoric about appropriate leadership that is articulated in operational environments is debunked using a continuum of leadership styles, from assertive to collaborative, as it relates to the level of risk and urgency.

'Hard' or 'soft' leadership may be required depending on the type of work being done and the risk profile of the systems of production used to do the work. The construction industry will be given as an example of the suitability of command-and-control project delivery. This will be supported by a discussion about some cultural factors that hinder risk-based decision-making including groupthink and stereotypes. A significant mindset transformation is overdue, especially for the OSH profession. Chapters 8 and 9 home in on being realistic about leadership.

There's no such thing as 'Safety Leadership', only 'Leadership'.

Otherwise, there also must be 'Environmental Leadership/Stewardship', 'Quality Leadership', 'Production Leadership' etc.

Apart from inadequate understanding of the nature of risk, the biggest gap in current risk management strategies is the lack of criteria on which to drive consistent risk-based decision-making at various levels of the organisational hierarchy. *PSM1* introduced the 'Reasonableness Test'. In Chapter 8, this will be adapted. It will describe a process that includes a Risk Ethos to test the suitability of decision choices and ensure the desired enterprise culture is sustained. Managers and supervisors should be taught how to apply the test in their work to balance short-term demands with longer-term objectives. This would shift timeframes, for instance, from achieving today's production target to continuity of safe production as an outcome of collective effort. The Reasonableness Test is specifically designed to eliminate unhealthy internal competition amongst managers and supervisors who are responsible for core business targets. This is achieved by ensuring that serious safety and maintenance issues aren't dismissed for perceived short-term production gains by one shift supervisor forcing the next shift supervisor to make the same decision. Will the 'weakest link' take responsibility for the escalating risks or will issues continue to be ignored?

The discussion will explain that the Risk Ethos is the backbone of the organisational culture defining how work is done towards common objectives. When supervisors make sound risk-based decisions, there should be greater consistency, confidence and earned credibility with peers and subordinates. Operational leadership needs to be adapted to situational variables and the current risk profile, whilst also being transparent and trustworthy. The latter can't be achieved if supervisors, because of production demands, compromise serious safety and/or maintenance risks. One poor decision can erode previously earned credibility.

The dialogue will turn to examples of how frontline supervisors can lead their team confidently through practical risk assessment exercises thereby earning trust and developing the competencies of their subordinates. There are issues within current OSH management systems that stifle practical approaches to risk management and prevent operational managers and supervisors from leading competently. Much of this stems from methods designed from an office perspective such as the use of spreadsheets or databases rather than something more meaningful to operational personnel, such as photographs and notes.

If head office needs collated risk information for business analytics, then the relevant function (OSH, quality, environment) should do this work rather than expecting operational managers to be masters of everything. They should be busy managing the operational risks not satisfying the 'system'.

Another important part of the operational manager's and supervisor's role is to keep their subordinates engaged with the work. The need for frontline leaders to effectively communicate with their team, particularly during onboarding of new personnel and whenever management of change is required, will be described. This is needed to prevent cultural degradation. Workplace conversations can revert to technical matters and social topics rather than supporting the team members' confidence and motivation especially through times of 'change fatigue'. This discussion will draw on the work of Grieve and Van der Stap who published, *'Safety and Entropy: A Leadership Issue'*, in the American Society of Safety Professionals' journal in 2020.

The chapter will close with an invitation for OSH Managers to scrutinise the management system to differentiate 'needs' and 'wants' with the view to 'right-sizing' the system and making it more inclusive of operational input. A clear explanation will be provided of how to shift from centralisation to decentralisation of the system whilst maintaining the capacity to deliver governance and assurance requirements for the Executive.

Chapter 9 builds on leadership development and turns to individual behaviours. People make decisions about whether to take risks or manage risks every day. This is normal behaviour. The issue at work is defining what level of risk, in other words, what are the boundaries between acceptable and unacceptable risks from an organisational (collective) perspective. The 'system' is supposed to define that but the intangible elements of that system such as the unspoken enterprise culture may be counteracting the overall strategy. It's important therefore to understand the relationship between risk-taking versus risk-managing and how these depend on decision-making processes, acceptance of behaviours, and the organisational culture. The Risk Behaviour Model (first published in 2008) will be updated to explain the interdependencies that connect individual behaviours, culture and learning outcomes within an operational context.

Fortunate technical, management and supervisory professionals get to work with a variety of different enterprises in diverse industries. This gives them exposure to the challenges of being engaged within divergent levels of maturity in terms of business risk appetite, production-centricity, and decision timeframes. Different strategies are required for these entities which for the purpose of the discussion will fall into the categories of 'The Good', 'The Bad' and 'The Simply Irresponsible'. At an organisation level, there can be distinct attitudes that determine risk-taking versus risk-managing as the preferred status quo. The new incumbent OSH or Operations Manager needs the competencies to readily identify where the enterprise sits on this continuum of non-compliance, to compliance, to beyond compliance because that will influence how they go about their work. Will they operate and maintain the current system with incremental adjustments or need a detailed gaps analysis to undertake a longer-term planned improvement strategy (the 'Good' and the 'Bad' respectively)? Have they walked into a business that needs a significant strategic overhaul including management's appetite for risk ('The Simply Irresponsible')? It should be borne in mind that 'compliance' isn't limited to OSH. It could include employment laws, taxation laws, company governance requirements, client contractual obligations etc.

These maturity categories differ at the fundamental level of organisational values. Unfortunately, personnel in operations seldom get the opportunity to consider and articulate their values and the impact these have on their behavioural decisions. These lay at the heart of developing risk management as a life skill. Too many training programmes are classroom-based and don't firstly, allow workers to develop practical in-field risk assessment competencies, and secondly, to understand their primary motivations for taking risks.

A key question is whether personnel are having an internal dialogue that prompts the question, 'Is it worth the risk?'

In Chapter 9, further practical strategies will be provided for supervisors to effectively lead day-to-day work and support subordinates to keep them engaged and practising risk management as a life skill. It's not an onerous strategy to increase the levels of operational participation and ownership of critical risk controls. Workplace Champions can be developed using a decentralised implementation plan by aligning critical control verifications with those who already have the subject matter expertise in the relevant area, for instance, electricians reporting on the operability of residual current devices. Others may be enlisted by providing essential training and development, for instance, fire wardens conducting drills but also verifying drill frequency against the plan for such activities and the follow-up on corrective actions.

In *PSM1*, there was extensive discussion on building organisational capacity. This was illustrated using the 'Capacity Reservoir' which will be adapted in this edition and shown in Figure 9.4. This is the collective competencies of the workforce for which there are numerous inputs. Capacity leads to an overflow of resourcefulness and resilience – qualities needed for continual improvement, optimisation of systems of production and the ability to cope with crises and threats. The process of building capacity will be illustrated beginning with an effective recruitment and selection process. Many inputs focus on enhancing human capital value such as accelerated learning using on-the-job competency development. Other areas requiring attention will be covered. These include right sizing the integrated management system, monitoring external and internal risks and opportunities, and developing a deep understanding of operations through data analytics to detect anomalies and trends.

Figure 9.4 which will illustrate the process of building organisational capacity includes expected outputs. These cover better review of systems of production, clear roles and accountability of risk owners, and a culture centred on loss prevention and exploitation of opportunities across multiple functions. Effective leadership continues as a key requirement for success. Further strategies will be provided to enhance the culture using affective, intellectual and action-oriented competencies. Throughout Chapters 8 and 9, the focus of professional development will be on managing risks and pragmatic, activity-based participation that enables managers and supervisors to lead in ways that are natural to the work, themselves and the team environment.

In Chapter 10, the discussions within previous chapters will be tied together, culminating in greater understanding of the principles applied by HROs. Aspiring enterprises can achieve business improvement be adopting these according to their unique circumstances. Some organisations, however, that considered themselves to be HROs, failed as evident by catastrophic events. This highlights the pitfalls of ignoring degraded systems of production and organisational factors such as cultural and leadership decay.

Operations need more practical approaches than simply following the principle of preoccupation with failure. They need to focus on day-to-day incremental losses that constitute waste. This ties back to the Entropy Model and the rising entropic risk curve with losses evident as production downtime, HSE incidents and quality deficiencies. These should draw management's attention, well ahead of escalation towards an inevitable catastrophic event, such as a fatality, force majeure, environmental disaster or loss of a key customer due to failure to meet quality requirements.

In Chapter 10, operations are provided with a reason to be sceptical about ALARP and the residual risk they inherit from design, construction and commissioning phases of their facility or site. Cost minimisation at the project front-end engineering and design phase can start a process by which sites inherent higher levels of residual risk and vulnerability to entropic risks (degradation) than appear on paper in the risk register. This warrants extreme or chronic wariness depending on the extent of HSE in design deficiencies within operation's infrastructure and technologies, and the processes by which these generate production. The discussion points to potential unknown unknowns (black swans) in the build that operations receive at handover after commissioning. New ways of looking for unknown unknowns are required. This may be aided by mathematical analysis of entropy and by challenging the accepted level of ALARP; effectively going back to the 'drawing board'.

Chapter 10 concludes by proposing ten strategies to move organisations towards high levels of reliability based on the Entropy Model and other tools discussed in this edition, in combination with adaptive HRO principles. These include the case for developing risk management as a life skill, which may be an opportunity to reduce the costs and training by silo-driven disciplines, where there's currently duplication and competing messaging, such as 'safety leadership' versus 'environmental stewardship'. This holistic approach should focus on building capacity towards multidisciplinary risk- and opportunity-based, enterprise leadership and culture development. Much of the waste in current organisations is a by-product of functional silos each driving their own agenda.

The ten strategies are diverse and will apply to various industries and enterprises in different ways. This means that critical thinking should be used when reading this edition. The final chapter consolidates some of the key themes developed throughout which include:

- honesty in assessing the identity of the enterprise;
- understanding risk not as a singular concept but involving residual and entropic risks that are dynamic, complex and ever-changing;
- caution in relation to current risk management practices and accepted bases for ALARP;
- better ways of managing risk day-to-day that lead to higher levels of participation, consultation and ownership; and
- having longer-term strategies to develop capacity and ways of working collaboratively.

A transformation from silo-based thinking to risk- and opportunity-based thinking is the critical mindset change needed for organisations to be successful in the Fourth Industrial Revolution. The reader, regardless of their professional background, is invited to make this transformation.

REFERENCES

BBC News, 2018, The Full Story of Thailand's Extraordinary Cave Rescue. *BBC News*.

Cambridge Dictionary, 2024a, IDEALIST | English Meaning. Cambridge Dictionary.

Cambridge Dictionary, 2024b, REALIST | English Meaning. Cambridge Dictionary.

Grieve, R., & Van der Stap, T., August 2020, Safety and entropy – a leadership issue. *Professional Safety Journal*, American Society of Safety Professionals, USA.

Hollnagel, 2021, *Synesis – The Unification of Productivity, Quality, Safety and Reliability*. London: Routledge - Taylor and Francis Group.

Klussmann, W., 2009, *Philosophy of Leadership – Driving Employee Engagement in Integrated Management Systems*. Hamburg: Diplomica Verlag.

Kymal, C., Gruska, G.F., & Read, R.D., 2015, *Integrated Management Systems: QMS, EMS and OHSMS, FSMS including Aerospace, Automotive and Food*. Wisconsin: ASQ Quality Press.

Law Insider, 2023, The Claimant Further Relied upon the Sample Clauses | Law Insider Barcock v Brighton Corporation [1949] KB 339.

Mol, T., 2003, *Productive Safety Management – A Strategic, Multidisciplinary Management System for Hazardous Industries that Ties Safety and Production Together*. Oxford: Butterworth-Heinemann.

Pardy, W.G., & Andrews, T., 2009, *Integrated Management Systems: Leading Strategies and Solutions*. Government Institutes, Lanham.

Sherf, E.N., Tangirala, S., & Venkataramani, V., 2019, *Research: Why Managers Ignore Employees' Ideas* (hbr.org)., Harvard Business Review.

Van der Stap, T., 2018, Chapter 29 Risk leadership – a multidisciplinary approach. *Safety Leadership and Professional Development*, American Society of Safety Professionals, USA.

1 The Entropy Model and Systems of Production

THE NATURE OF RISK

Over the years, safety culture branding has morphed into mythology and rhetoric. This is despite many supervisors and workers struggling to make heartfelt commitments to concepts such as 'Zero Harm' or 'Safety First'. Why? Realistically, there's no such thing as zero incidents or zero injuries simply because risk can't be eliminated. The straightforward task of pruning the roses is an example. It's a good outcome to complete the job without a scratch but chances are that despite gloves and a long-sleeved shirt, there will be at least one minor injury as the day's work goes on. Likewise, it's easy for 'Safety First' to be followed when there's no production pressure. Once deadlines or targets are looming, the incentive to expedite the work by taking shortcuts increases. Hollnagel referred to this as the Efficiency-Thoroughness Trade-Off (Hollnagel, 2023).

The gap between 'zero' and 'ouch' is probably as wide as WAI and WAD. The managerial motivation for investing in branding probably came from well-meant intentions to articulate commitment to the health and safety of workers. There is to a degree a logical argument behind this; however, it's the operational implementation that lacks credibility. Specifically, the stance that is taken to 'Zero Harm' is that if we, as an organisation, don't espouse zero, then we must be saying that injuries are acceptable. In the workplace, it's difficult to explain to workers that 'zero' is an ideal or aspirational goal because workers think practically not idealistically. Examples such as this will be expanded in Chapter 2 which discusses the 'Alignment Fallacy' – why safety policy and strategy doesn't consistently become operational reality, especially in the context of strong core business drivers.

To challenge the delusion of zero, *PSM1* presented the Entropy Model to redefine risk. This will be illustrated later in this chapter. Risk management practices continue to rely solely on reducing residual risk (the remaining risk after all controls have been implemented). The Entropy Model, however, introduced the concept of 'degradation' of systems with a resulting rising 'Entropic Risk'. From the outset, the model proposed:

> *There's no such condition as 'zero' risk so therefore 'Zero Harm' is a fallacy as is nil incidents and injuries. Debate about 'Zero', philosophically or statistically, is noise that the original Entropy Model (Mol, 2003) sought to silence.*

Many companies embraced branding and some companies continue to advertise vacancies for a 'Zero Harm Manager'. It would be a brave incumbent who puts hand

DOI: 10.1201/9781032701561-2

to heart to that commitment! Imagine making the claim that my role is to deliver on our Zero Harm Policy. Step one: Remove all roses (and other thorny matters) from the site.

The original Entropy Model in *PSM1* was republished by the American Society of Safety Professionals in 2018 in the book, *Safety Leadership and Professional Development*, "Chapter 29 – Risk Leadership: A Multidisciplinary Approach". This chapter provides a more detailed description of the phases of the original model through a series of illustrations that didn't appear in *PSM1*. In this edition, there are also further iterations of the model to provide different methods to which it may be applied.

WHAT IS 'ENTROPIC RISK'?

'Entropy' was originally a term used in thermodynamics which explains the tendency of the system to move to a state of disorder or randomness. In *PSM1*, this was defined more clearly as systems degradation with these systems being inputs to the production process. These are the process (how the work is done), technology (plant, equipment etc. used to undertake the work), the physical environment (the workplace where the work is done) and human resources (the person/people doing the work). Each of these factors tends to degrade resulting in a rise in the level of risk, referred to as 'Entropic Risk'. Systems shift to a state of loss of control or chaos.

Since *PSM1*, this concept has been widely accepted. For instance, Hollnagel (2021) supports the concept explaining:

> The purpose of controlling a system can be seen as an attempt to reduce entropy, which means to reduce disorder and lack of predictability.

Further:

> Waste can be seen as entropy, as in Lean. Lack of quality can be seen as entropy, as in statistical quality control or Kaizen. Accidents can be seen as entropy, as in Safety-I.
>
> *(Hollnagel, 2021)*

Hollnagel also discusses "anthropogenic entropy" which refers to

> the increasing disorder and decreasing predictability that are due to our fragmented understanding of the world and hence the inadequate steps that are taken to restore order and the inaccuracy of the interventions that are made. In the case of socio-technical systems, the surroundings – the external forces – are themselves socio-technical systems. These are never stable or constant but always develop and change for better or worse. Anthropogenic entropy to a large extent exists because of the psychological reasons for fragmentation. Because we can never completely comprehend the whole situation, and because of the limitations in how far we can look ahead and how thorough we can be in considering alternatives, there is a need for constant adjustments and corrections that, because they are always approximate, become a source of entropy.

The concept of entropy has also been adopted in security management. Interestingly, security and OSH have many parallels including the primary objective of preventing losses or failures by identifying the risks, assessing those risks, and implementing adequate controls. Such sufficiency is characterised by depth of defences whereby multiple controls are implemented to prevent failure. A study by Coole (2010) found that:

> within a systems approach to physical security there is a complex interrelationship between the built environment, physical controls, technology, people and management processes as they achieve the elements of Defence in Depth. Within this complex inter-relationship the study indicated that decay occurs at the constituent level, and if left undetected expands to affect the sub-system in which it is located......
>
> The study suggested that security decay theory is primarily concerned with managing the natural entropic processes/pressures occurring against commissioned levels of effectiveness within PPS. In addition, the study indicated that in order to maintain PPS at their commissioned levels of effectiveness during their life cycle, they need to be managed in accordance with their commissioned designed specifications.

*PPS is Physical Protection System

There are two notable learnings from the study. The first is the acknowledgement of the tendency of systems to degrade and the level of risk rising accordingly. The second is the inference that there is an optimal level at which to maintain the system (in this case the PPS) and this is according to an agreed design. Coole and Brooks (2014) paper, "Do Security Systems Fail Because of Entropy?" expanded on the earlier concepts. In summary, they found:

> Security is implemented to mitigate an organisation's identified risks, linking layered elements into a system to provide countermeasure by the functions of deter, detect, delay, response and recovery. For a system to maintain its effectiveness these functions must be efficaciously performed in order; however, such systems may be prone to decay leading to security failures.

Their publication describes the emergence of entropy to many domains including information security, organisational systems, combat systems, communications and other fields. All of these discuss the degradation and disorder within a system in relation to a system's capacity to undertake work.

In the following section, the Entropy Model is explained which applies this concept of entropy to risk and resultant losses including production downtime, OSH and environmental incidents, and quality deficiencies. Many applications of entropy as described above can be readily absorbed at the managerial and professional levels where strategic and operational thinking are inherent in the language and deliverables of this type of work. But:

It's insufficient for managers and professionals to understand the nature of risk. It must be understood by everyone which means that any 'model' must be explainable in terms of real-world operations.

This is a major differentiator of the Entropy Loss Causation Model.

It readily transfers concepts to concrete applications.

THE ENTROPY MODEL: A SYSTEMS APPROACH

The Entropy Model presents two types of risk – residual risk and entropic risk. Everything humankind does has some risk associated with it. In business, energy is used to source product, to transform, transport and consume. The amount of energy involved is the inherent risk associated with the process that is being undertaken. By establishing controls, this inherent risk is reduced to as low as reasonably practical. For instance, when driving at 60 miles per hour, the energy level is still present but by designing and manufacturing vehicles with safety features, the residual risk is reduced. By design, society can lower residual risk through technological improvements, which generally occurs over the longer-term through innovation and economies of scale as these safety features become more affordable and standardised.

The Entropy Model is intended to be a diagrammatic rather than mathematical representation of the nature of risk. The left-hand column in Figure 1.1 shows the 'perfect system' where work processes, technology, the physical environment and people are the inputs of the organisation. The level of risk is zero as shown by the solid line and perfect safety, production output and quality are achieved. This part of the model is the unreachable ideal. The practical reality is that a level of residual risk (shown as the solid block in the next column) is always present. This can't be reduced in the short term due to technological, financial and resource constraints. The probability of an incident or loss is never zero. In hazardous industries, such as mining, oil and gas, transportation and manufacturing, this residual risk is relatively high compared to non-energy dependent businesses, and therefore, there is always potential for losses.

The Entropy Model identifies that residual risk is inherent in the four systems of production that combine when a task is undertaken, and collectively, when an organisation carries out production. These systems are processes, technologies, the physical environment and human resources. In addition, these systems degrade over time, sometimes gradually and at other times rapidly. This is referred to as 'Entropic Risk' which rises concurrently as systems degrade, as shown in Figure 1.2.

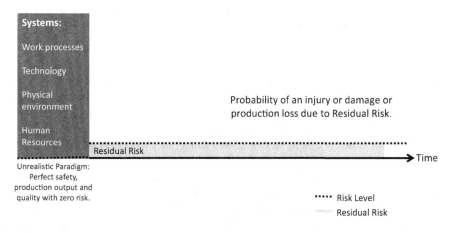

FIGURE 1.1 The first section of the Entropy Model

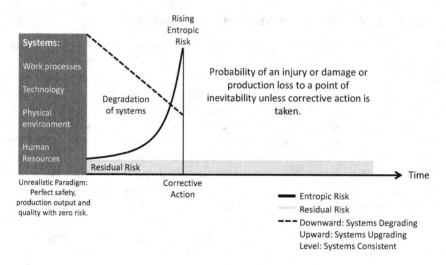

FIGURE 1.2   The second section of the Entropy Model

As a point of interest, in more recent times, a different discipline called 'Fuzzy Mathematics' developed and used Shannon's entropy approach with the correlation to safety and risk exposure lending weight to entropy as a category of risk. This work was done independently of the Entropy Model in *PSM1*. These contemporary research studies support the concept of 'entropy' as a risk factor, if not a risk type, for instance, Larranaga (2013) refers to this concept as 'Disaster Creep' and states:

> Many modern energy, industrial, transportation, health care, telecommunications, and political systems are highly vulnerable to small changes that propagate and develop into major disasters. This systemic tendency to unravel, decay uncontrollably, or move from order to disorder (e.g. disaster) is a characteristic of all natural and human-made systems and is defined by Newton's Second Law of thermodynamics as 'entropy'. Before one can overcome this entropic tendency of all systems to unravel, one must understand the nature of system failure across multiple disciplines.

(Larranaga, 2013)

Referring to Figure 1.2, the four systems of production degrade as shown by the dashed downward line, and concurrently, the probability of an incident or loss to the organisation increases, shown by the rising entropic risk curve. The four systems can degrade singularly (e.g. human fatigue), concurrently (e.g. technological and physical environment decay due to lack of maintenance) or collectively across all four at varying rates. This rise could have been shown as a straight line – a direct proportion to the rate of degradation. Instead it's shown exponentially to illustrate the concept of rapid potential shift to chaos and inevitable losses, as was characterised by some catastrophes such as the Challenger Disaster of 1986.

As a practical example, a food processing company had an inadequately managed, paper-based systems, aging infrastructure and equipment, poor leadership commitment to compliance and low operational competencies amongst supervisors and workers. Overall, the systems were in a constant state of degradation. Chaos was the status-quo and tolerated by management. Safety performance was characterised by a 'revolving door' of musculoskeletal injuries, longer-term repetitive strains and incidents ordinarily associated with 'fatal risks' such as falls from heights and electric shocks.

Whilst the safety-related incidents were high so were personnel issues, environmental incidents, quality deficiencies, re-work, absenteeism, and internal conflict.

Across all functions, the evidence of 'entropic risk' was tangible and alarming.

Many incidents caused by entropic risk result in no injuries and may, if personnel are in the vicinity of the failure (in the 'line of fire'), be classified as near misses. Such events are a concern for safety but also other functions. For instance, production will be impacted by downtime, maintenance by increased reactive repairs, environmental management by spills and potential regulatory breaches. The model indicates that OSH incidents are an outcome of this escalating risk; however, these are only one category of loss. Other sub-standard performances or losses can include the failure to consistently achieve production targets, required quality standards, reputational damage etc. (For the sake of simplicity, only injury, damage and production loss are shown in the figure.)

Figure 1.2 explains what needs to be done when escalating entropic risk is identified due to degradation of one or more systems of production. The first action is to determine the degradation source/s and take corrective action to prevent further decline. Such action circumvents potential incidents or losses. The failure to do so can result in continued rise in entropic risk to a point where negative consequences become inevitability and potentially catastrophic.

At the strategic level, corrective actions can be responses to warning signs. These may relate to production, quality, safety and/or other business objective. For instance, threats to the supply chain may compromise one or more of these functions as many companies found out during COVID-19. At the site level, safety incidents may be considered triggers for raised concern particularly if these lead to halts to production or draw the attention of the regulator. The level of response required can range from local issues management to crisis management depending on the overall impact to the organisation.

It's important to consider evidence of entropic risk and its impact from a strategic perspective and to ask, 'What's going on in this enterprise overall?'. Does management have a clear understanding and control of the risks? Is there sufficient discipline at the site level where the organisation has its greatest operational risk exposure?

Referring again to Figure 1.2, corrective action, both strategically and operationally, is critical at any point along the degradation continuum. When companies go through a cycle of warning signs, quick fixes, operate as normal, warning signs, quick fixes, it's often because of the failure to recognise degradation and take corrective action until the undesirable outcome becomes absolutely apparent. The most obvious and alarming of these conditions is when a workplace fatality or other catastrophic event occurs. Workplace OSH performance is usually measured using lagging indicators such as lost time injury rates, but this may be looking at safety through too narrow a lens.

> *A series of serious safety incidents should raise senior management's concerns that there is loss of operational control more broadly across functions not just safety. That's because entropic risk is a collective state of systems degradation that doesn't just affect safety. It affects everything.*

Chronic wariness should trigger interrogation of what's working, what's not working and a heightened precautionary approach to present and emerging risks. Questions to ask include: Does senior management have line-of-sight on operational risk management? How is it measured? How is it reported? Most importantly, does it trigger the right level of corrective action based on the risks to production, safety, quality and other business objectives? Management should be increasingly wary of potential vulnerabilities, seek to understand these, and act on them with a sense of urgency. This is shown as the vertical line at which corrective action is taken in the model that prevents further entropic risk.

The Model continues to be developed in Figure 1.3. This illustrates that proactive maintenance practices are required in all systems of production to prevent future degradation. System standards are raised, as shown by the upward dashed line, whilst concurrently the probability of incidents and other types of failures is lowered to the residual risk level. This is shown by the downward solid line representing

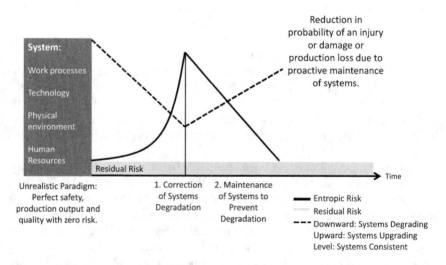

FIGURE 1.3   The third section of the Entropy Model

decreasing entropic risk. Where it meets the residual risk level is the lowest point that can be achieved in the short-term due to resource, financial and technological constraints. Organisations don't have unlimited money to reduce residual risk and therefore this must be done through planned investment as part of business improvement budgeting.

The Entropy Model is readily taken from concept to practical application. It's been explained to managers, supervisors, workers and in community forums open to small-medium business owners through to corporate leaders. It's easy to understand when framed in a practical way. For instance, a person's hobby shed at home can be used to visualise how it applies. When a group of people is asked about the states of their sheds at home, there is invariably at least one person who confesses to having a mess. They give examples such as poor housekeeping – lack of orderly storage, disused equipment and materials and time wasted finding the right tools for the job. When the 'confessor' is asked to share a reflection on their efficiency, personal safety and the impact on the quality of their work, their story is often described in good humour and delivers a clear message about room for improvement.

When the scenario is expanded to a workplace where productivity is expected and success as a team depends on the standard of everyone's inputs, it becomes evident that entropic risk is counter-productive to workplace, business unit and organisational success. People recognise the need for corrective actions but also that they shouldn't accept repeated, reactive quick fixes at work as the solution. There also needs to be preventative systems in place to avoid recurrence (of the mess). This involves practices that maintain standards, which is the maintenance phase shown in Figure 1.3. Implementation of this phase requires the support of operational managers and supervisors who provide the resources to improve these systems of production. At the most rudimentary level, there needs to be a place for everything and everything in its place.

The 'system' developed and implemented through consultation and participation drives the right behaviours, manages risks effectively and provides the opportunity to pursue production, safety, and quality work concurrently.

Storytelling along these lines applies to every home and workplace, across cultural and language barriers, with only the need for local adaptation. The principles apply everywhere to everyone because they reflect how life works. The first three parts of the Entropy Model make sense to managers and workers alike. It's when the rest of the model is presented that the 'big picture' emerges. This has the power to transform behaviours and bring about positive change in a short period of time.

As an example, an exploration company had a maintenance workshop provided by the owners but most workers using the facility were independent tradespersons. Each had their own specialisation such as heavy diesel mechanics, light vehicle mechanics and boilermakers. As a result, the workshop had been divided into specific trades areas, however, there was often conflict over housekeeping in shared spaces. Frustrations regularly boiled over and were exacerbated when crews came back from the field and dumped equipment needing repair anywhere. This created a distinct 'them' (the field crew) and 'us' (the service crew). Management considered this normal in this 'tough' industry and believed it would resolve itself without their intervention.

At the regular operations and safety meetings, the question was consistently asked, 'What cheeses you off?' (an Australian colloquial way of asking what can be improved). 'What's holding us back from doing our best work?' With the collective desire for change evident through these open discussions, basic risk management workshops were held using the Entropy Model as the conversation point. The story-telling of the home shed was used to get each small group in the right frame of mind and to point out the practicalities of basic operational risk management and owner-ship. These were some of the changes made with management's support:

- Each trade was given the authority to set up their work area to enable safe, efficient, quality work;
- One senior tradesperson (also a contractor) who worked most effectively with everyone was asked to be the workshop supervisor. This person became the 'go-to' person for suggestions and further overall improvements that were taken to management for approval;
- The risk team (HSE and operational risk) assisted by ensuring compliances were in place such as emergency response equipment, signage and other vital support requirements for the facility as a whole;
- The field crew were given a designated laydown area for equipment that needed repair (which stopped the dumping in the workshop). They were also shown how to lodge maintenance requests using a mobile phone-based operations app;
- A person was employed to undertake the daily housekeeping so that the tradespersons could focus on the high value work of maintaining and repairing plant and support equipment essential for continuity of production and safe operations.

The risk workshops empowered the contractors and gave them authority to make the changes they needed. Without further instruction or direction being given, the results were:

- Each area ended up with a place for everything and everything in its place;
- The work became more efficient, safer and to higher standards of output quality;
- Conflict was reduced significantly;
- Tradespersons were learning from each other, in fact, there was healthy competition about whose area was the best;
- The barriers between field crew and maintenance reduced;
- Continuous improvement became embedded in how the business worked;
- Operational discipline became self-driven and if anyone's standards slipped, the team used positive peer pressure to keep standards high.

This became their new reality of work. In the past, time was wasted looking for parts including nuts and bolts. Within a matter of months, the workplace was orderly, structured and an operational system (both safe and efficient) was set up by the people who did the work. This fine-tuned to the extent that trays were mounted along the wall that contained all the different types of screws, nuts and bolts. The trays were labelled with the supplier's barcodes so that the supplier's sales representative could come in, do the stock check, and organise resupply of sparing quantities. That's efficiency and reliability in action. This is a small-scale example but a significant one to that business. Even large organisations can be divided into smaller operational areas to enable similar outcomes. The example shows how top-down and bottom-up strategies can work concurrently, which is a topic of discussion in later chapters.

So far, only the first sections of the Entropy Model have been explained. It's when the whole model is described that people start to understand why corrective action and maintenance practices are critical and why these need to be sustained. Summing up the discussion so far, the root cause of entropic risk is the tendency of systems of production to shift towards a state of chaos. Proactive maintenance is the counteraction to this chaotic effect. Corrective action is reactive whilst proactive maintenance practices are a managed response to recurrent issues and known risks. Both are required and set the standards of performance by which an enterprise operates strategically and at the individual worker level. From a management perspective, monitoring practices and data analysis must accompany the strategy. A sustained wariness of potential failure is needed to trigger timely corrective action with a burning concern for safety, efficiency and quality performance. The 'safety' lens alone is too narrow to achieve such outcomes.

'Safety' doesn't exist in isolation from other business objectives, especially production.

If managers and supervisors when doing their walkarounds notice poor housekeeping standards, then there's nothing to be gained by blaming the workers if the fundamentals of safety and efficiency in design aren't in place. Telling the workers to take corrective action to create some order won't prevent recurrence. The key questions to ask are: What systems need to be in place to encourage the desired housekeeping standards? What changes need to be made to set a higher standard of safety, quality work and efficiency? Importantly, how can we best get the people who do the work involved in any modifications so that the change is supported and sustained?

As the model evolves, Figure 1.4 illustrates the level of risk that a company should be targeting. This effectively is ALARP. Degradation is prevented by reactive corrective action as needed and implementation of proactive maintenance practices mitigates the risk of future degradation. The company achieves optimisation when the risk level is brought down to the residual risk level. Accordingly, a compliance outcome is also achieved.

The accepted industry definition of ALARP according to the Health and Safety Executive (UK) is:

'ALARP' is short for 'as low as reasonably practicable'. 'SFAIRP' is short for 'so far as is reasonably practicable'. The two terms mean essentially the same thing and at their core is the concept of 'reasonably practicable'; this involves weighing a risk against the trouble, time and money needed to control it. Thus, ALARP describes the level to which we expect to see workplace risks controlled.

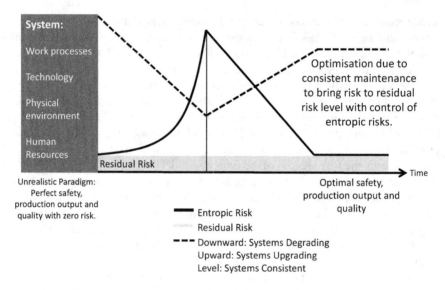

FIGURE 1.4   The fourth section of the Entropy Model

Using 'reasonably practicable' allows us to set goals for duty-holders, rather than being prescriptive. This flexibility is a great advantage. It allows duty-holders to choose the method that is best for them and so it supports innovation, but it has its drawbacks, too. Deciding whether a risk is ALARP can be challenging because it requires duty-holders and us to exercise judgment.

(Health and Safety Executive, 2024)

When considering the residual risk level in the Entropy Model, ALARP shouldn't rely on lower order controls such as administration and personal protective equipment (PPE). This is because these mitigations are not robust or consistently reliable because they depend on human performance which is inherently fallible.

Due to entropic risk, ALARP should rely on higher order controls in the Hierarchy of Controls being elimination, substitution and engineering controls, as the lower order controls of administration and PPE are vulnerable to degradation.

The discussion about ALARP and its deficiencies will continue later in Chapter 3. Returning to the model, there's no room for complacency in the organisation. Figure 1.5 provides a reminder that systems tend to degrade so if decision-making and behaviours don't sustain the focus on risk management, systems will go into decline again and the probability of incidents and failures will escalate.

There's sound statistical evidence to support this cyclical nature of organisational performance. For instance, a review was undertaken of fatalities in the state of Queensland in Australia from 2000 to 2019, referred to as the Brady Review.

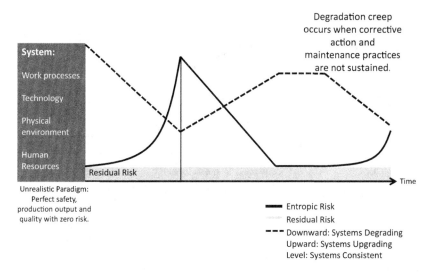

FIGURE 1.5   The fifth section of the Entropy Model

The Brady Review found that there is a fatality cycle evident in the industry – meaning, there are periods when fatalities occur, followed by periods when there are few to none... An explanation for this is that when a significant number of fatalities occur, industry tightens up safety requirements in response; however, this may then be followed by a drift into complacency and failure where the fatality rate increases. This can also be demonstrated in the fatality cycle, in the financial year of 2004–05 where there were four fatalities, but in the following financial year there were half the number of fatalities (two) and then in 2006–07 there were four fatalities. A similar cyclical pattern continues in the available data.

(Resources Safety and Health Queensland, 2022)

THE FOUR STEPS RISK MANAGEMENT STRATEGY

The Entropy Model has practical application to any workplace and provides clarification of what needs to be done to manage the risks of work. The model drives effective risk management to pursue safety, efficiency and quality concurrently.

Each function's performance is an outcome not an end in isolation from other functions. Risk and opportunity management is the overarching strategy upon which all performance outcomes depend.

The model guides the behaviours of managers, supervisors and workers. The frontline supervisor should be the Risk Champion coaching the workforce to promote corrective action and proactive maintenance as an organisational discipline. These two steps are shown in the model in Figure 1.6. These are the first steps in the 'Four Step Risk Management Strategy', which aims to counter complacency about entropic risk and instil the required operational behaviours.

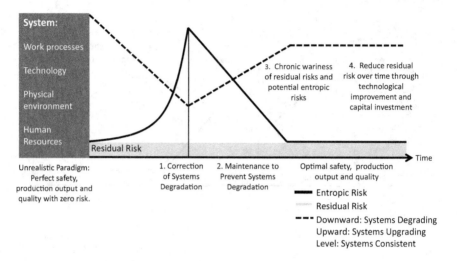

FIGURE 1.6   Another version of the Entropy Model explaining four steps for effective risk management on a timeline

This is important because systems can be perceived to be so safe in some highly regulated, high-risk workplaces because of the depth of risk mitigations, to the extent that the workforce is lulled into a false sense of security. Residual risks associated with energies are contained, guarded, isolated or controlled in such a way to prevent the worker from being 'in the line of fire'. It's assumed that personnel can concentrate on the task at hand alert only to more minor hazards. The modern, seemingly well-controlled workplace allows complacency to creep in despite residual risk. This is also evident within society, for instance, in relation to driving of motor vehicles. Whilst safety in design has improved significantly over the last 50 years, respect for the energies involved has diminished. This manifests as risk-taking choices including intoxication, fatigue, distraction and speeding.

Complacency results from the failure to acknowledge that entropic risk is real.

This complacency is evident in post-incident language; 'you were lucky' 'to get out of harm's way' 'in the nick of time'. Luck, evasion and timing are readily identified as 'life savers'. Steps 1 and 2, corrective action and preventative maintenance of systems of production, should be designed to take the luck, line of fire and bad timing out of the equation.

The Entropy Model shows residual risk as consistent because it can't be reduced in the short-term. Figure 1.6 provides step 3 indicating that the organisation and workers must remain alert. The higher the total residual risk of the four systems in combination, the greater the need for vigilance and operational discipline in implementing required controls. Examples include heavy vehicle road transport in hazardous conditions, remote oil and gas platforms, and underground mining operations with difficult geotechnical conditions. These are high-residual risk industries that require rigorous controls to prevent losses (including catastrophes).

The level of residual risk which is acceptable is a strategic decision driven by cost-benefit analysis and availability of alternatives. The front-end of a project where design, construction and commissioning decisions are made will determine the level of residual risk inherent during the operation's lifecycle. The Model shows that with step 4, residual risk may be reduced in the longer term through technological innovation and economies of scale.

PSM1 explained in detail the Four Step Risk Management Strategy. This is a practical tool for both management decision-making and individual behaviours. Summarising, the steps are:

1. Take immediate corrective action to eliminate entropic risk;
2. Establish maintenance strategies to prevent future entropic risk;
3. Manage residual risk in the short-term by remaining alert and exercising operational discipline when implemented required controls;
4. Reduce residual risk in the longer-term through investment, technological innovations, and economies of scale.

The Four Step Risk Management Strategy is a life skill that defines workplace practices that are much more meaningful than simply the instruction to comply with the management system.
These apply at work and at home.

THE TOTAL RISK PROFILE

The two types of risk found in each of the four systems of production combine when work is done. As a result, a comprehensive risk profile can be developed through scenario development. A detailed breakdown of risks was provided in *PSM1* with a chapter dedicated to each of the four systems. Table 1.1 provides a summary of the risk interfaces or sources and indicative controls. The need for a multidisciplinary approach is evident when looking at the detail. There are inputs from engineering (e.g. technological design), OSH (e.g. major hazard standards), production (e.g. site design and layout), human resources (e.g. values-based recruitment), environmental management (e.g. site environmental studies), information technology (e.g. performance monitoring and big data), maintenance (planned replacements) and training (e.g. pre-operational training).

Total risk profiling in this way highlights the interdependencies that exist within an organisation. For instance, the rate of wear and tear of truck components will be affected by the behaviour and competencies of the operator from the production department. The mechanical reliability of the truck will be affected by the behaviour and competencies of the tradespersons from the maintenance department. The two departments are therefore responsible for preventing technology-related entropic risk. For this reason and because of the complexity of potential system interfaces, a multi-disciplinary approach to risk management is a must. Functional silos should be broken down by applying a unifying risk-based strategy across the business.

TABLE 1.1

Residual and Entropic Risks and Indicative Controls by System of Production

System of Production	Residual Risk	Entropic Risk
Processes: Lead to the combination of technologies with the physical environment and people (human resources) Controls for all interfaces through process safety and risk assessments of all phases of process lifecycle and management of change.	Process/technology interface Process/physical environment interface Process/human resource interface	Process/technology interface Process/physical environment interface Process/human resource interface
Technology: Include all man-made plant, equipment, tools and enabling information technologies	Design and manufacturing shortcomings • Pre-purchase evaluation • Supplier information • Performance monitoring and big data Technology/physical environment interface • Design and planning of work environment • Hazop studies • Installation standards • Major hazard standards Technology/process interface • Standardisation of technologies • Risk assessment Technology/human resources interface • Ergonomic design • Job design • Task modification	Wear and tear • Proactive scheduled maintenance • Inspection and monitoring regime • Planned replacements • Reactive maintenance Technology/operator interface • Pre-operational training • Organisational culture and desired behaviours • Correction of undesirable behaviours
Physical Environment: Include both manmade infrastructure and natural environments which can constitute a place of work (whether temporary or permanent)	Site design shortcomings • Site environmental studies • Documentation of residual risks Physical environment/process interface • Site assessment and selection • Site design and layout Physical environment/technology interface • Design and planning of workplace • Pre-installation risk assessment • Safe work practices	Natural degradation • Regular monitoring • Emergency response and crisis management • Maintenance regime Physical environment/process interface • Monitoring and inspection regime • Safe work practices • Proactive scheduled maintenance • Reactive maintenance

(Continued)

TABLE 1.1 (CONTINUED)

Residual and Entropic Risks and Indicative Controls by System of Production

System of Production	Residual Risk	Entropic Risk
	Physical environment/ human resources interface • Modifications to fit worker • Flexible work practices • Monitoring of worker health • Residual awareness training	Physical environment/ technology interface • Monitoring and inspection regime • Proactive scheduled maintenance • Reactive maintenance Physical environment/ human resources interface • Monitoring for person fit • Inductions and training • Housekeeping • Culture of desired behaviours
Human Resources: The human capital utilised to undertake the work.	Incomplete competencies • Inductions and training ongoing • Risk awareness training ongoing • Change management • Communication systems	The new employee • Value based recruitment • Mentoring • Supervision • Hazard monitoring (person is hazard)
	Physical limitations • Physical environment modifications • Matching capability to task • Sick leave entitlements • Job rotation and enrichment • Safe work procedures • Preventative health measures	Fatigue • Workplace, technology and process modifications • Job rotation • Fatigue monitoring • Training Ill health • Training and self- assessment • Healthy lifestyle programs • Monitoring of worker wellbeing Stress • Education and training • Employee assistance programs • Dispute resolution • Right sizing of staffing levels Safety violations and drug & alcohol breaches • Organisational culture • Disciplinary procedures • Risk identification and management training • Communication processes • Pre-employment testing • Random testing • Healthy lifestyle programs • Employee assistance programs

Substandard work practices and a culture of complacency contribute to production losses, quality deficiencies and increased risk of HSE incidents. The Four Step Risk Management Strategy derived from the model is the socio-technical and behavioural driver for operational discipline and a risk-based culture.

Explaining the Entropy Model with practical examples helps personnel to understand why the quality of their work is critical to overall risk management.

THE CYCLICAL RISK OF FAILURE FOLLOWING SUCCESS

Figure 1.5 of the model showed a potential cycle of degradation, corrective action and maintenance, optimisation of sustainable organisational performance and the tendency to shift to degradation again. This can be referred to as phases of failing, mending, optimising and recurrent failing, which aligns to the discussion above about operational performance. Figure 1.1 illustrated the non-existent, perfect world where risk is absent. Figure 1.2 introduced the degradation of systems of production leading to rising entropic risk and inevitable, tangible failure if not addressed. Figure 1.3 showed what may occur post-incident or loss. If an organisation isn't ordinarily proactive, the enterprise having learned the hard way may react by mending 'broken systems' through maintenance practices. Where such practices become robust, the organisation achieves optimal safety, continuity of production and quality outputs in Figure 1.4. Complacency over time then results in degradation again as shown in Figure 1.5 and so the cycle is repeated.

Hazardous industries and those that experience production/ productivity/ schedule/ other core business pressures need a better understanding of risk than traditional safety accident causation models provide. The way forward is a whole-of-business approach rather than the simplistic OSH focus that defines operational risk as the presence of hazards. A hazard is a set of circumstances that may cause harmful outcomes and the probability of it doing so, coupled with the severity of the harm, is the risk associated with it. In the siloed OSH field, it's the likelihood of loss evident as injuries that matters. This is a limited view of what constitutes loss for an organisation. In many cases, safety incidents are accompanied by other damages such as production downtime, indirect costs of incident investigations and harm to workforce morale and company reputation.

The focus should always be on 'Good Work' which is safe, efficient, and set by clear expectations regarding the quality of the output.

In the workplace, and in fact the wider community, the presence of risk isn't considered a cause for concern. It's the degree of risk that matters. Similarly, the perception of 'safety' also centres on the level of threat to people. What 'safe' means to most is that the risk associated with a particular activity is 'negligible' and to make something sufficiently safe means to reduce the risk to an 'acceptable' level (British Medical Association, 1987). The gap is that risk, from a strategic perspective, affects business performance more broadly and as discussed later, there are potentially flawed beliefs about what's an acceptable level of risk. Entropic risk isn't part of operational risk language, but it should be.

The traditional safety-based accident causation models summarised in *PSM1* have a strong emphasis on the role of human error rather than identifying human resources as a system of production. Human capital should be developed more objectively using a risk-based approach to build capacity. The question is how should HR residual risks be managed and HR entropic risks prevented? The revolving door of train and re-train has missed the point. Personnel should have risk management skills that are enduring, transferrable and continually improving through experience. Currently, having provided training after an incident to close competency gaps, a delusion of safety can result. Management tends to believe that once something has been done about a hazard or people's skills to manage the hazard, it's now 'safe' or 'safe enough'.

The post-incident re-training corrective action is too often centred on relearning procedural controls rather than upskilling risk management competencies such as planning, re-assessing conditions during task execution, effective communication, and reinstating the workplace to the required standard after the work is done. Where the tasks have been high risk, a review of what worked well and where improvements can be made should also be part of closing the loop to embed effective risk management processes and competencies. Other safety-related models don't explain what workers are practically expected to do to manage risk and this is a significant gap that the Four Step Risk Management Strategy closes.

Unfortunately, the perception from traditional safety models that all accidents are preventable if human error is eliminated is counter-intuitive to the concept of building organisational capacity. Archaic models have created a narrow view focused on negligence, for example:

> Workplace accidents are caused by people. More accurately, they are caused by the things they do or do not do. Equipment and machinery will sometimes fail, and incidents may occur which cause accidents, but they are nearly always traceable to some degree of human error, negligence or ignorance.
>
> *(Robert-Phelps, 1999)*

This mindset can still be found in organisations that are low on the maturity journey. It encourages a blame mentality. Some managers have said, "I don't want to go to jail because of some idiot!" (The 'idiot' being a worker.) This attitude is destructive for the organisation and its people, especially when Operations and OSH Managers are trying to build a supportive risk-based culture. A blame approach can run concurrently with a fallacious belief that having a paperwork system is compliance, rather than understanding use of the paperwork to capture actions and evidence those actions to inform further risk-based decision-making.

One of the major gaps in organisations whose hierarchy wants to blame the workers for incidents and other losses, is a lack of understanding of their role as managers and leaders.

The day-to-day dialogue of OSH Professionals must counter blame
but not at the expense of accountability.

A good starting point to assist these managers and supervisors to move away from blame is to expose them to a systems approach driven by objective risk and opportunity management. Within academic literature, some traction has been achieved to challenge the blame culture but operationally, there's a long way to go. The shift away from simplistic human error as the cause of safety incidents has built momentum. For instance, Conklin writes that people make mistakes and errors are normal. This is completely true when we objectively consider our own behaviour, for instance, is there a driver who has never made a mistake or nearly caused a collision? The discussion is important to frankly accept the fallibility of human beings within the broader system. These are the fundamental principles about error currently accepted by the OSH profession that Conklin (2019) proposes. Firstly, error is seen as normal and never causal. It's also not the opposite of success because success can occur despite errors. Sometimes errors lead to failure but not always. People don't commit error intentionally, so error is only a result of a behavioural choice in retrospect. Human error can't be removed so it must be planned for and accepted as inevitable. Good systems build in error tolerance. Conklin also suggests that error without serious repercussions should be used in leading indicator data analysis.

Human error should be recognised as a HR risk within the context of the systems of production. In practical terms, this has OSH, productivity and quality performance implications. As suggested above, good systems should build in error tolerance with the inference that consequences should be reduced but can't be entirely avoided. Since *PSM1* which focused primarily on OSH, production and quality outcomes, Hollnagel has recently added 'reliability' to the suite. There is a wider acceptance of the need for a multidisciplinary approach which Hollnagel called 'Synesis' (Hollnagel, 2021).

Here is the opportunity.

OSH Professionals, having made the transformation themselves, should be
influencing managers and supervisors to shift their mindset from 'safety
based' to 'risk based' thinking focused on the reliability
of systems of production.

Reliability involves safe production delivered to expectation quality
standards.

The narrow approach to worker behaviour (unsafe acts) and work environment (unsafe conditions) is redundant. This is an important step forward in both a practical and an ideological sense – acknowledging that no one wants to be injured at work and we rarely choose to or knowingly act in a dangerous manner of our own volition at work. 'Unsafe acts' are usually symptoms of systemic problems such as inadequate knowledge of the risks involved, work pressures, excessive physical or psychological demands associated with the task or conditions, or an unhealthy organisational culture.

The focus on human error is a hindrance from exploring fully the underlying variables that lead to losses. These are often attributable to inadequate or degraded organisational systems.

The role of human error in accident causation is by no means excluded from the risk profiling provided by the Entropy Model. All system of production including HR, have a residual risk due to limited capacity in terms of knowledge, skills and abilities (KSAs) and also physical constraints. No individual has complete competencies to anticipate and deal with every activity or to solve every potential problem that may arise in the workplace. As listed in Table 1.1, HR also suffer from degradation leading to rising entropic risk. This can affect any individual worker or group of workers, such as the short-term risks of fatigue and ill-health and in the longer-term, the aging workforce.

PSM1 was concerned with risks associated with systems of production. In more recent years, organisational factors – less tangible elements of the enterprise – have been given attention especially as a result of the findings of many major disasters. Several of these have uncovered flawed decision-making within unhealthy cultures as a root cause of catastrophe. The examples below include the Challenger and Columbia space shuttle disasters. The system failures associated with these catastrophes

were largely due to the focus on technology, lack of consideration of the social and political system within NASA, and inattentiveness to the consequences of organizational structure and policies on human behavior and safety (Vaughan, 1996; CAIB, 2003: Chap. 8). The [Challenger] Rogers Commission Report also identified "flawed decision making" as a contributing cause of the accident, in addition to other factors such as production and schedule pressures, and violation of internal rules and procedures to launch on time (Rogers et al., 1986). Ethical issues raised in the Challenger disaster involve engineering responsibility versus management decision-making, as well as the ethics of post hoc whistleblowing and negligence in design. Vaughan (1996) characterizes the Challenger launch decision as a case of 'organizational deviance' – values and norms that are seen as unethical outside the organization but are seen within as normal and legitimate. Flawed decision-making was a common factor in both of these accidents, however, the decision-makers saw their actions as rational (Leveson, 2011). Leveson argues that enhancing the cultural and organizational environments and appropriate tools are needed to improve decision-making under conditions of uncertainty.

The crash of two 737 MAX passenger aircraft in late 2018 and early 2019 resulted in the death of 347 passengers and crew. While there were significant concerns about Boeing's organizational culture, flawed design choices were a key factor that led to these crashes. Boeing's management placed a higher emphasis on cost reduction, profits and stock price at the expense of safety (Robinson, 2021). The Federal Aviation Agency (FAA) regulatory body gave 'self-certification' authority to Boeing which conflicted with passenger safety. Engineers were also responsible for making decisions that involved trade-offs between cost and safety, however, the culture and Boeing's management policy on minimizing costs and adhering to a delivery schedule discouraged engineers to dissent. Boeing engineers and managers violated widely accepted ethical norms such as informed consent and the precautionary principle (Herbert et al., 2020). A key theme with these major incidents is the conflict over production (or its equivalent such as schedule) and safety.

(Van der Stap and Qureshi, 2024)

Incident investigations such as these point to organisational factors being root causes of devastating events. The author classified these factors and presented them in a paper at the American Society of Safety Professionals Conference in 2019. The proposed areas under 'Organisational Capacity' were leadership, competencies, management systems and resourcefulness with resilience. These were contextualised within the Entropy Model as having both residual risks and vulnerability to degradation. This revised version of the Entropy Model will be presented in Chapter 8 to consider systems of production risks along with organisational factor risks.

SUMMARY

This chapter has delivered a challenge to industry's current understanding of the nature of risk using the Entropy Model. The failure to recognise that systems of production (processes, technology, the physical environment and human resources) tend to degrade is a significant gap in knowledge and operational implementation that needs to be overcome to prevent catastrophic events but also day-to-day losses.

'Entropy' is an important concept, not only in OSH but also in other functions. It originally came from thermodynamics but has been more widely applied to the tendency of systems to shift to a state of disorder or chaos. A primary organisational objective is to prevent disorder and maintain control to achieve objectives. The Entropy Model (first published in *PSM1, Mol 2003*) shows that as systems degrade, there is a resulting rising 'entropic risk' which causes losses in terms of safety incidents, production shortfalls, quality deficiencies and other hindrances to enterprise objectives.

The model includes the residual risks of systems of production. These are shown to be at a consistent total level in the short-term due to the organisation's resourcing constraints. This is also affected by the availability of technological solutions for risk reduction. Better HSE in design options don't become accessible to enterprises until the costs of such innovations come down to the extent that the organisation can justify the outlay through cost-benefit analysis.

In this chapter, the eight sources of risk were explained. These are the residual and entropic risks in each of the four systems of production. The complexity of these eight sources and the interfaces between them means that a multidisciplinary approach to risk management is required. The systems approach version of the model can be applied both strategically and operationally. As a learning tool, it can be readily understood by managers and workers alike using practical scenarios. This is a major step forward in enabling a deep insight into the nature of risk, which is currently too readily bundled into a singular, over-simplified concept. Risk is complex and dynamic.

A systems approach is required to break down silos that exist according to functional areas in organisations. A common direction and way of work must be driven by effective risk management. The Four Step Risk Management Strategy from the model provides a framework for this to occur. Every function has a vested interest in loss prevention. Production departments suffer from downtime. Human Resources is concerned with turnover, absenteeism and poor morale. Maintenance aim to minimise breakdowns and rework. OSH aim to prevent incidents and injuries. All losses result from risk which is the underlying threat to functional performance.

All organisations should aim to optimise their systems of production to avoid losses. The model illustrated this as a phase where such systems are raised to a standard of quality where degradation is controlled, and the risk level is brought down to the residual risk that can't be reduced further in the short term. This provides industry with clarity about what constitutes effective risk management.

This is accompanied with a warning. Organisations have the potential for failure following success if corrective actions and maintenance practices aren't sustained across the systems of production. The history of many industries shows this trend. Following a significant incident, efforts are heightened to regain control and investments are made into systems improvements. With time, complacency sets in, and systems are allowed to degrade again, for instance, because of deep cost cutting. Eventually, losses start to rise such as near-misses or production downtime. These may be ignored or put down to bad luck. Left uncorrected, systems degradation and accompanying rising entropic risk will lead to another serious incident or disaster, and so the cycle of failure following success is repeated.

In Chapter 2, the discussion will turn to the reasons why organisations haven't to date matured towards a risk-based approach. A major factor is areas of non-alignment in current management systems that cause potential conflict between safety and production. Three levels of alignment will be explained to contextualise risks and opportunities within current enterprises. Management attitudes to compliance can be a major stumbling block to business improvement so the discussion will cover exploiting OSH legislation to strategise for better safety, efficiency and quality outcomes concurrently. As an organisation matures, it should consider the benefits of attaining ISO accreditation or alternatively, self-assessment including building awareness of the growing ESG Agenda in society.

REFERENCES

British Medical Association, 1987, Living with Risk. Chichester: John Wiley and Sons. Quotation from p.56.15.

CAIB (2003). Columbia Accident Investigation Board Report, Volume *I*. Washington, DC: Columbia Accident Investigation Board.

Conklin, T., 2019, *The 5 Principles of Human Performance*. Independently published, USA

Coole, 2010, *The Theory of Entropic Security Decay: The Gradual Degradation in Effectiveness of Commissioned Security Systems*. Edith Cowan University.

Coole and Brooks, 2014, Do security systems fail because of entropy? *Journal of Physical Security*, 7(2), 50–76.

Health and Safety Executive (UK), 2024, ALARP "at a glance", Risk management: expert guidance. ALARP at a glance (hse.gov.uk)

Herbert, J., Borenstein, J., & Miller, K. (2020). The Boeing 737 MAX: Lessons for Engineering Ethics. *Science and Engineering Ethics*, *26*(6), 2957–2974. https://doi.org/10.1007/s11948-020-00252-y

Hollnagel, 2021, *Synesis – The Unification of Productivity, Quality, Safety and Reliability*. London: Routledge - Taylor and Francis Group.

Hollnagel, 2023, ETTO principle (erikhollnagel.com).

Larranaga, 2013, *Disaster Creep – Why Disaster and Catastrophe are the Norm – Not the Exception*. Fire Protection Engineering, Issue 71: Disaster Creep – Why Disaster and Catastrophe are the Norm - Not the Exception – SFPE.

Leveson, N. (2011). Engineering a safer world: systems thinking applied to safety. Cambridge, MA: The MIT Press.

Mol, T., 2003, *Productive Safety Management – A Strategic, Multidisciplinary Management System for Hazardous Industries that Ties Safety and Production Together.* Oxford: Butterworth-Heinemann.

Resources Safety and Health Queensland, 2022, Facilitating high reliability organisation behaviours in Queensland's resources sector and modernising regulatory enforcement. Consultation Regulatory Impact Statement, FINAL-CRIS-Version.pdf (rshq.qld.gov.au).

Robert-Phelps, G., 1999, *Safety for Managers - A Gower Health and Safety Workbook.* Hampshire: Gower Publishing Limited. Quotation from p. 29.

Robinson, P. (2021). *Flying Blind: The 737 MAX Tragedy and the Fall of Boeing.* New York: Doubleday.

Rogers, W.P., Armstrong, N.A., Acheson, D., Covert, E.E., Feynman, R.P., Hotz, R.B., Kutyna, D.J., Ride, S.K., Rummel, R.W., & Sutter, J.F. (1986). *Report of the Presidential Commission on the Space Shuttle Challenger Accident.* NASA.

Van der Stap, 2018, Chapter 29 Risk leadership – a multidisciplinary approach. *Safety Leadership and Professional Development*, American Society of Safety Professionals, USA.

Van der Stap, T., & Qureshi, Z., 2024, *Can Organizations Achieve High Reliability and Manage Risk More Effectively Using the Entropy Model?*, Unpublished work at the time of writing.

Vaughan, D. (1996). The Challenger Launch Decision: Risky Technology, Culture, and Deviance at NASA. Chicago: University of Chicago Press.

2 The Path to Organisational Maturity

THE ALIGNMENT FALLACY

'Safety First' can't consistently be practised at the operational level when core business (production, productivity, schedule, stakeholder satisfaction etc.) pressures are present. It could be argued that such pressure is itself, a 'hazard' – not of a physical but of a psychosocial kind. It doesn't matter who a person is, everyone has failed to put 'Safety First' at some time when doing a task. People may consciously rush (at work or at home), for instance, by speeding to an appointment or by not returning to the garden shed to get their gloves later to be injured on a rose bush.

Organisationally, the phenomenon of 'drift to danger' under pressure was presented by Rasmussen in 1997. He said that there is

> systemic migration of organizational behavior towards accident under the influence of pressure towards cost-effectiveness in an aggressive, competing environment
>
> *(Rasmussen, 1997)*

Whether an organisation is for-profit or service-driven, management accountability for performance will, in some way or another, apply tension towards increasing efficiency. It's challenging at best and more realistically, contradictory to brand the enterprise as a place where safety is advocated above all other functions. 'Safety First' falls into the same category as 'Zero Harm' – a slogan to promote an ideal. At a toolbox talk, the oil and gas general manager may address the workforce and ask for 'Safety First' but in the next breath highlight that production is down and everyone needs to work harder and/or faster to get the next shipment out to the customer. In the health care sector, the hospital director may also talk about 'Safety First' in relation to staff safety but then talk about the decline in patient care based on patient related incidents or complaints. Patient safety will likely be perceived as taking precedent over staff safety.

'Safety First' is an ideal that can never consistently survive the operational reality of work.

In organisations there are mixed messages about safety versus production and this was described in *PSM1* as 'The Alignment Fallacy'. Figure 2.1 shows, on the left-hand side, the presumptions about alignment where business policy, goals and values translate into strategic management decisions that cascade to middle- and lower-management decisions before being implemented operationally. It's assumed

DOI: 10.1201/9781032701561-3

FIGURE 2.1   The alignment fallacy

that the enterprise's goals and values and employees' goals and values are aligned and that there's no noise or disruption within this top-down process.

Over the years, particularly in immature organisations, this approach has been perpetuated by a mindset that command and control is the most effective means of generating business performance. This hierarchical strategy fails to consider several issues. The first is that the higher a manager is on the corporate ladder, the further they are from actual control and the 'coalface' where the operational risks materialise. The second is the perceived need to be saying the right thing, which without a sound understanding of the operational challenges, can come across as rhetoric.

It's tempting to shroud OSH performance expectations in 'Safety First' that appears to uphold high moral values, but the critical piece is whether operational personnel believe it in the context of their operational reality.

It's not unusual for operations to feel a disconnection from head office. Sometimes board members and executives prefer not to know the harsh truths of the business even when cost-effective solutions are being presented to them. If production targets are being achieved, how this is done, including what compromises must be made, can be irrelevant to those who are far removed from the shop floor. As an example, a meat processing company had a production line where the downstream workers had no control over the pace of work. As a result, raw product often fell off the conveyor belt onto the floor (which was a breach of quality control and food hygiene laws) and created hazardous conditions as workers moved around the factory. One of the workers incurred a serious facial injury from his own knife and quickly moved away from the production line holding his face in his hands. The other workers didn't leave their station on the processing line to help. The incident was captured on CCTV and reviewed by management who adopted a 'blame the worker' attitude. The video and investigation findings were presented to the board who had insisted historically on branding the management system 'Safety First'. The OSH manager pointed out the sheer impossibility of actualising this culturally in an environment where production targets are the only priority.

'Safety First' is a lie when workers have no control over the pace of production.

Hopefully, this example is rare and at the extreme end of organisational culture. Ironically, businesses that are highly production-centred, immature and risk tolerant can suffer significant losses as a daily occurrence. These can include safety, production, quality deficiencies and personnel issues; however, the appetite of these businesses for change can remain tepid at best. In all organisations, material risks are most prevalent at the operational level. This is where the work can go well or very badly. 'Safety First' won't resonate with frontline supervisors and workers if management neglect to provide a safe workplace. These are matters of significant non-alignment.

There are numerous factors that contribute to the failure of policy to cleanly cascade to operations. This is illustrated in Figure 2.1, on the right-hand side, as the Alignment Fallacy in Practice. The dilution starts with strategic management decisions and become more pronounced according to the disconnection between strategy and practice. This is particularly the case in production-driven companies simply because core business is why the organisation exists in the first place. As a result, the tension between production (go faster and generate more) and safety is perpetuated. Hollnagel (2009) described this as the Efficiency Thoroughness Trade-Off (ETTO). To be more efficient, attention to detail is compromised.

This tension between production and safety applies in many organisations. When thinking about 'production', the industries that come to mind tend to be oil and gas, mining, manufacturing, agribusiness and the like. 'Production' however, has an equivalent. The key consideration is 'What is the source of pressure?' For instance, the pressure in restaurants is time-related to ensure customer satisfaction. In hospitals, it's associated with ensuring the standard of patient care and in some cases, the safety of the patient takes precedent over that of the worker. Would a nurse allow an elderly person to fall to the ground so they themselves don't end up with a manual handling injury? In schools, teachers prioritise the quality of teaching and student care. This pressure may lead to mounting stress that affects the health of the educator.

Accounting practices work against safety policy and loss prevention management systems more generally. There's a financial conflict between profit and the costs of managing risk plus it's impossible to account for all losses from poor risk management or the full benefits of effective risk management. The problem with grandiose labels for safety such as "Safety First" is there's no clear limit of how safe the work must be to be considered safe enough to proceed. It can be a weapon for dispute when management and workers disagree on the level and adequacy of risk mitigation measures. "It must be made safer. Why? 'Safety First'".

Management systems can contain weaknesses in terms of non-alignment with the 'Safety First' message. Examples include reward systems based on production bonuses or individual performance evaluations that are output-focused without adding weight to managing risk for loss prevention. For instance, at a mine where the workers were undergoing risk management skills training, one of the issues that came up was that workers were ignoring the faulty park brakes of the four-wheel drive vehicles that were taken underground. They had a pre-start checklist to complete and were falsifying the condition of the park brake marking it satisfactory instead of honestly reporting faults. Their motivation was that the vehicles (most of which had this fault) would have been sent to maintenance. This downtime would have led to less production. They believed that targets would suffer and so would their production bonus payments.

The challenge with turning around this collective, non-compliant behaviour was to help workers understand timeframes and consequences of risk-taking including personal loss. It was also to awaken them to the goal of consistency and continuity of production rather than today's production. There was an underlying distrust of management. Workers believed that reporting wouldn't result in change. In fact, the workers assumed that management knew of the vehicle maintenance issues and blamed them. When it was explained that management weren't aware because the pre-start forms said everything was satisfactory, they realised that the issues wouldn't change unless they started to complete the checklists honestly.

It's a mistake to assume that senior managers know what the issues are especially if they aren't receiving truthful information from the operational level. Managers simply can't dedicate resources to problems they don't know about.

As shown in the right-hand side of Figure 2.1, the site level is where the conflict between safety and production is most evident. This is particularly the case in the role of the frontline supervisor who's concurrently responsible for safety and production within management systems that don't necessarily support both roles simultaneously. For example, if a shift supervisor fails to achieve production by fixing the safety and maintenance issues that have arisen on their shift and previous shifts, they're accountable. If production is pushed to the detriment of safety and an incident occurs, they're also held accountable.

The frontline supervisor faces a dilemma which is inherent in the role and not well supported by management in core-business-centric cultures. There may also be unspoken internal competition between production supervisors to achieve the highest output targets. This may be occurring to the extent that safety and maintenance issues are ignored by some supervisors on their shift, later to be rectified by the supervisor who's more conscientious about safety and reliability.

As explained above in the example of underground workers falsifying vehicle pre-start checklists, at the operational level, employee goals and values can be non-aligned with company goals and values (shown at the base of the right-hand side of Figure 2.1). This is particularly the case when there's a culture of 'them and us'. Management blames the workers for underperformance of the business whilst workers blame management for not listening, not acting, and not seeking to understand their day-to-day challenges. There can be a history of false assumptions from both parties towards the other. This can be left unaddressed causing an undercurrent of covert conflict.

The consequences of the alignment fallacy include poor risk management, production centricity, poor decision-making, lack of effective leadership and absence of heartfelt commitment to health, safety and wellbeing. With cultural symbols and management systems in place, the hierarchy can develop the illusion of control. Management can become attracted to this false impression of structural mass control through uniform bureaucratic processes which then seem to make their roles as managers and supervisors much simpler and less messy (Tate, 2013). 'Safety First' is an example of this over-simplification.

In theory, 'Safety First' consistently drives operations.

The continued occurrence of fatalities, serious injuries and debilitating illnesses suggests otherwise. Managers and supervisors need better leadership support, coaching and decision-making tools to achieve safe, efficient, quality work.

Assumptions of control are also evident at the operational level. In the ASSP article (Grieve and Van der Stap, 2020) the cycle of failing, mending, optimising and failing (adapted from the Entropy Model in Figure 1.5 of Chapter 1) was applied to individual worker performance and the influence of supervisors. This version of the model is presented in Figure 2.2.

Because people are inherently fallible, a low level of HR residual risk must be accepted within the 'perfect system' as shown on the left-hand side. The never-ending cycles of degradation, recovery, optimisation, degradation, recovery etc. are phases shown progressively in the figure. Phase 1 is the ideally designed system: a system without emotion, history or interpretation. Phase 2 is the initial degradation which begins immediately because of the gap between the worker's knowledge, skills, and

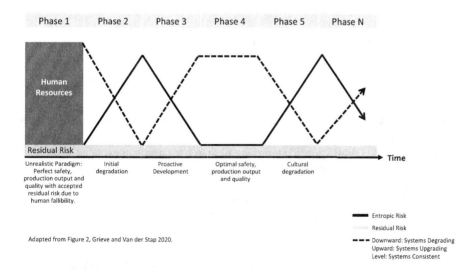

FIGURE 2.2 The entropic risk in the human resources system of production

abilities (KSAs) and those required to fit their new employer organisation. In Phase 3, there is recovery from degradation to a period of optimisation (Phase 4) where individual-organisation fit is achieved in terms of competencies to an adequate performance level. This can be followed by Phase 5 where cultural degradation may occur because of lack of alignment on underlying goals and values as organisational truths are revealed during employment.

Many people start a new job with great optimism to find with time, that most organisations face similar issues, such as conflict, time pressures and other negative aspects of the psychological contract.

After Phase 5, the cyclical nature of the process is recognised by entering Phase N. This pattern is due to the tendency of the People (HR) system – an input of the whole system – to degrade. The decline is shown as the dashed line whilst the associated entropic risk is illustrated with the solid line that rises and falls over and above the residual risk level. The solutions proposed by Grieve and Van der Stap are discussed later in this book when supervisory strategies to prevent HR entropic risk are covered in Chapter 9.

The causes of non-alignment between potentially conflicting objectives as well as corporate policy versus operational reality requires a multidisciplinary review. This isn't something that an OSH professional or any single functional silo can address. A collective leadership strategy is required, and executive management must give permission to the leadership team allowing them to ask difficult questions. The Alignment Fallacy is an invitation to critique the contradictions that exist in the organisation, especially where there are potential areas of functional conflict in operations, such as safety versus production.

There's no such thing as the perfect system so there will always be tension particularly given the need for tough decisions in the face of limited resources. The opportunity is to understand these non-alignment issues and to explore areas for improvement. The fundamental questions are:

- How can managers build credibility through better messaging about risk (rather than 'beating the safety drum')?
- What can be done differently to support the role of the frontline supervisor who's responsible for safety and production concurrently?

It would be easy to point the finger of blame at the safety function. Where gaps such as these are found with the benefit of hindsight or the guidance of academic thought leaders in the OSH profession, it's tempting to search for a logical explanation of how things came about. For instance, who came up with 'Safety First'? Why did the managers at the time buy into this doctrine? Why has the OSH profession shifted from and between 'Safety-I', 'Safety-II', 'Safety Differently' etc? Operationally, the likely response to these alternative paradigms would be indifference.

'The Work' has always been what matters in operations, not safety branding, symbols, or paradigms. Operations need solutions not academic debate.

Some of the fundamental issues of non-alignment arise before a venture even gets off the ground. As an example, a company was setting up a new project and had recruited a core team of managers. They were seated around a table for the first time and after introductions, the conversation shifted to how to deliver this project differently. A key area of discussion was the espoused culture. Each manager provided their thoughts as follows.

Construction Manager:

We can deliver a world-class project efficiently, like my last project – low-cost country sourcing and modularisation. It's fast and keeps workforce numbers to a minimum. We don't want any schedule issues.

Quality Manager:

Quality assurance is the key to a successful project. I'll be ensuring our suppliers know what's expected of them. There won't be any compromise on material standards and the quality culture will drive construction.

Environment Manager:

We have a clean slate with this project and community expectations are high about environmental issues. The regulators are going to be very interested in how we manage impacts. Our commitment and culture need to be visible to stakeholders.

OSH Manager:

My priority is to get good OSH systems in place and to build a strong safety culture from day one. We need management's commitment and leadership, especially during construction when our site risks will be high.

Each manager put forward their own functional priorities advocating a discipline-based culture. Who was right? Who would 'triumph'? Would schedule pressure during construction make these words meaningless?

The example highlights that silos are currently inherent in how organisations work. The interdependencies and interfaces based on risk and opportunity haven't broken this mould and little progress has been made towards common drivers, despite some obvious synergies. For instance, when *PSM1* was published in 2003, OSH and HRM were distinctly different functions. In 2024, the overlaps are unquestionably accepted as including training and development, leadership, culture, capacity building, resilience and other human performance programmes.

From a practical perspective, the synergies between OSH and operational risk management are strong but unfortunately, in many companies under exploited. As a result, there's operational work and safety work as two distinct elements of the job; the safety paperwork needs to be done before the job can start. Recent literature includes some pieces on the integration of operational management with OSH (Hasle et al., 2021) who state:

> Increasingly, large and medium-sized enterprises rely on joint management systems to align safety and health processes with operational issues such as productivity, quality, and environmental management ... The decoupling of OSH management and operations management has for many years been conceptualised from the OSH perspective as the 'sidecar' problem ... where OSH management is a separate activity, organised in its own logic.

The issues are known but the literature is lacking in terms of delivering the know-how to integrate OSH and operations. This is a key challenge progressively addressed in this book. The starting point is to better understand the context of the organisation and the issues that affect all functional areas, regardless of whether the enterprise is a large corporation or an SME.

THE THREE LEVELS OF ALIGNMENT

In *PSM1*, it was explained that the organisation doesn't exist in isolation from the outside world. For it to succeed and endure, it must create fit with its environment and adapt when changes occur, effectively maintaining 'external alignment'. This concept was not entirely new in 2003 when *PSM1* was published. The basic principles could be found in books about strategic management. A key theme throughout is, "Why is external alignment important?"

Most large corporations understand why and many of the requirements are embedded in corporation law and accountabilities to shareholders and other stakeholders. For the SME owner or manager, this discussion may seem remote and perhaps mute in relation to day-to-day operations, but change is inevitable and affects everyone, particularly in the Western and Developing Worlds. For instance, the impact of technological changes requires responsiveness to maintain relevance in an increasingly competitive and aggressive marketplace. SMEs (not just big corporations) benefit from lean, operationally driven, risk-based management systems. In some ways, SMEs have advantages over larger organisations in their ability to make changes quickly and seamlessly with production. In fact, those companies that elect to have their systems tested through International Organization for Standardization (ISO) accreditation, may gain a competitive advantage when bidding for contracts. They may also improve the businesses' reputations with clients and community.

For this reason, the Strategic Alignment Channel from *PSM1* is reintroduced in this edition. It's shown in Figure 2.3 and applied using practical examples for various sized businesses. The term 'channel' is used to imply unobstructed flow through the enterprise where it adapts to upstream risks and opportunities from external sources. For instance, most organisations were required by law and necessity to adapt to COVID-19 restrictions mandated by relevant regulators.

The governing and influencing forces shown in the figure in the left-hand column include changes in legislation, technological innovations and other dynamics. As a further example, the full impact of climate change strategy is yet to cascade to SMEs and households but the likelihood of expanding requirements in almost guaranteed. Impacts are already being identified as described on the World Economic Forum website (World Economic Forum, 2022).

Looking at the figure, the 'flow' extends to operations because they have upstream risks and opportunities from the strategic level of the organisation. Internally, these may include an organisational restructure, introduction of new goods or services, or increased accountability of sites to head office requirements. For instance, OSH Corporate may develop standards for the prevention of fatalities and serious injuries, for operational implementation. These are often referred to as 'fatal risk controls' or 'major hazard standards' (examples can be found on the internet). In mining operations for example, there are often common causes of deaths and injuries including:

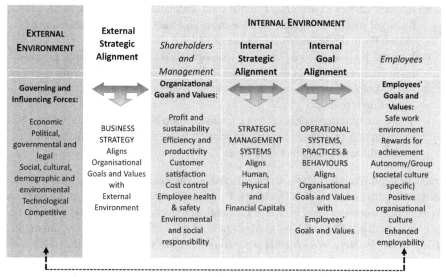

EXTERNAL ENVIRONMENT	External Strategic Alignment	INTERNAL ENVIRONMENT			
		Shareholders and Management	**Internal Strategic Alignment**	**Internal Goal Alignment**	*Employees*
Governing and Influencing Forces:		Organizational Goals and Values:			Employees' Goals and Values:
Economic Political, governmental and legal Social, cultural, demographic and environmental Technological Competitive	BUSINESS STRATEGY Aligns Organisational Goals and Values with External Environment	Profit and sustainability Efficiency and productivity Customer satisfaction Cost control Employee health & safety Environmental and social responsibility	STRATEGIC MANAGEMENT SYSTEMS Aligns Human, Physical and Financial Capitals	OPERATIONAL SYSTEMS, PRACTICES & BEHAVIOURS Aligns Organisational Goals and Values with Employees' Goals and Values	Safe work environment Rewards for achievement Autonomy/Group (societal culture specific) Positive organisational culture Enhanced employability

Interdependencies and Feedback

FIGURE 2.3 The strategic alignment channel

vehicles, surface, and underground mobile plant operations; unstable ground conditions; hazardous and molten materials; moving equipment parts; exposure to energised sources; and falls from heights. Operations managers and supervisors should be involved in determining the critical controls required to prevent fatalities in their workplaces and the programme for monitoring effectiveness. On face value, SMEs may feel that these 'fatal risk controls' don't apply to them, but many of the same inherent dangers are found in SMEs as in large organisations.

> *There's a universal suite of energies and conditions that can result in workplace deaths. Not all will apply in every workplace, but no business is exempt from fatality potential.*

A safe work environment is an entitlement and recognised as such in Figure 2.3 under 'Employees' Goals and Values' in the right-hand column. Other employee goals and values may vary between countries or industries, but the basic common needs of workers are listed as shown - a safe work environment, rewards for achievement, autonomy/group (societal culture specific), positive organisational culture and enhanced employability.

The key elements of the Channel are:

- External environmental forces drive business strategy which then cascaded to strategic management systems and onto operational systems, practices and behaviours;
- When business strategy aligns to the external environment, 'external strategic alignment' is achieved, which assists the enterprise to remain relevant and meet its organisational goals and values, such as profit, customer satisfaction etc as listed;

- Internal strategic alignment centres on balancing human, physical and financial capital to support core business and the support services required internally to deliver required outcomes;
- Internal goal alignment requires that operational systems, practices and behaviours embed alignment of organisational goals and values with those of the employees, which is necessary for retention and return on human capital investment;
- There are interdependencies and a feedback loop between the external environment and employees because of social values, which don't remain static, each influencing the other over time.

Practical examples are helpful to explain these concepts. In 1998, the International Labour Organization made a Declaration on Fundamental Principles and Rights at Work which were amended in 2022. This included everyone's (globally) entitlement to a safe and healthy working environment (International Labour Organization, 2022). In the Western World, OSH legislation has been improved significantly since the 1990s to ensure that statutes are written in such a way that new and emerging risks are captured through 'enabling' legislation. In simple terms, 'enabling' means the employer has a 'duty of care' to workers for their health and safety. Recently in Australia the definitions of 'employer' and 'worker' have been expanded to capture more circumstances where a person is working regardless of whether they're paid.

Such a significant change in global social values has translated to local OSH legislation to which no enterprise is exempt. Any business owner who elects to ignore these requirements is essentially accepting a substantial business and personal risk (due to penalties that apply to enterprise 'officers') should they fail to fulfil their duty. This example shows the connection between the 'big picture' (global level) to the 'small picture' (local level where an SME employs a few workers).

This isn't scaremongering but a reality check on the dynamic nature of risk and opportunity within the business world. In fact it would be safe to say that when the original channel model was presented in *PSM1* in 2003, the gap between the 'External Environment' and SME strategies was chasmic. Today however, because of the worldwide web and availability of information, that gap has narrowed significantly. Other policy issues such as climate change, the gig economy and workplace psychosocial health and wellbeing are rapidly gaining momentum. For instance, small businesses that are suppliers to some major Australian mining companies are required to submit a declaration of compliance to that company's modern slavery policy, to remain preferred suppliers. Can a small business owner in Australia be certain that there was no child labour used in the manufacturing of the personal protective equipment they provide to their personnel, when such products are made in Asia? Matters such as these present a real dilemma for SME owners and managers as client expectations increase into realms over which these owners and managers have no control. What are they to do – sign the declaration or risk losing a customer?

Importantly, regardless of organisational size, external environment factors don't affect a single function such as OSH management or production. This overarching 'umbrella' of risk and opportunity filters through to most if not all functions.

The following is another practical example taken from the COVID-19 pandemic. Prior to the COVID-19 outbreak in 2019, many organisations had a crisis management plan within which 'pandemic' was a small section – included mostly because it was felt by managers that it should be because it was industry practice. The scope of this was perceived to be mostly associated with sustaining workforce levels in case of an influenza outbreak. During the COVID-19 pandemic, because of directives of various government health departments, regulatory bodies and / or corporate head office instruction, some organisations undertook risk assessments and implemented response plans. The complexity of these plans was highly dependent on the nature of the industry and external factors such as that industry's supply chain.

For instance, in an agribusiness that has pork production as core business, the risk assessment can be complex and wide reaching. A narrowly focused OSH risk assessment based on workforce health would be grossly inadequate. The risk assessment method would require review of inputs, processes, and outputs. For example, the following inputs would be critical. For a given farming capacity of 30,000 pigs, production doesn't cease, and thousands of pigs are sent to the abattoirs every week that are soon replaced by the planned next generation. If pigs can't be sent to the abattoir, there's no spare capacity to sustain them on the farms. Pigs can't be released into a paddock and kept there until the pandemic is under better control because they're not herd animals and adopt cannibalism behaviours.

Critical process issues could include how soon to start slaughtering animals and disposing of carcasses; the health and environmental issues associated with burial to landfill; availability of suitable landfill sites given travel restrictions associated with the pandemic; the response from the public and from animal rights activists, and how this should be managed.

In the event of a case of COVID-19 at an abattoir, the logistical challenges begin. This includes how to stop or divert trucks full of pigs already on the way to the abattoir; how to dispose of production output that is already being processed (meat, carcasses and waste) and then undertake a deep clean of the facility; how to manage personnel who often socialise outside work; how to plan return to work scheduled for future normal operations; and recovery of stock number at the piggeries given the need for strictly planned fertilisation, gestation and growth durations.

Output issues could include abattoir waste disposal at downstream rendering plants; what to do with waste if the rendering plant also must be closed; where to send the waste and how to dispose of it without creating environmental impact or community nuisance; and the need to meet customer contracts and consumer demand. Other concerns include disruption to food supplies and preventing panic buying by the public.

Examples such as this highlight the need for a multidisciplinary approach to risk management. If the assessment method solely looks at functional areas, for instance, only the production line or worker health and safety or quality, the major issues won't be identified. External threats can create complex, inter-related interfaces within the organisation and often solutions need to involve external stakeholders such as regulators, government, contractors and customers.

The value of the Channel is to clearly map and understand internal processes, external influencing forces and stakeholder needs, within the context of normal operations. This means not waiting for crisis conditions to prompt such analysis.

This also means building relationships with stakeholders as a strategic response to the dynamic environment that defines the context of the organisation.

INTEGRATING ISO STANDARDS

The Channel illustrates the overarching framework for an integrated management system – one, for instance, in which operations and OSH can be seamless. It should be noted, that where legislation states that the organisation must have a 'safety management system' (SMS), there's nothing to prevent the business from integrating SMS requirements into their operational plans, procedures, and work instructions. The SMS doesn't have to be a separate, standalone system. The key is to understand what OSH legislation has prescribed (the must-do's). Taking this further, when 'safety issues' arise, these can be resolved by also considering efficiency gains. For instance:

Plant not in use
> The person with management or control of plant at a workplace must ensure, so far as is reasonably practicable, that plant that is not in use is left in a state that does not create a risk to the health or safety of any person.

(Government of Western Australia, 2022)

To meet this legislation, compliance may be as straightforward as ensuring that keys aren't left in the forklift when not in use. Keys should be returned to the workshop office and kept there. The efficiency gain is that when the forklift operators need to use a forklift, all the keys are in a known place, not left on a bench or in someone's pocket. The instruction can be part of the operations checklist rather than a separate safety instruction. Although this is a very basic example, it's surprising how much time is wasted in workplaces simply because there isn't 'A place for everything and everything in its place'.

One of the benefits of attaining ISO accreditation is that understanding legislative requirements is a mandatory element. Other requirements that lead to a structured, systematic approach to the work include having: an applicable policy; clear roles and responsibilities; processes for participation and consultation with workers; planning for risks and opportunities; clear objectives; adequate resourcing; competency and awareness development; processes for communication; control of documented information; operational control across multiple risk areas; internal monitoring; management review and continual improvement (International Organization for Standardization, 2024). These are areas which have commonality with operations.

A systematic approach can be achieved by adopting ISO Standards or by exploiting synergies across different functional areas as explained by the International Organization for Standardization (2024). For instance, loss prevention applies to multiple departmental areas such as production (downtime), OSH (fatalities, injuries, illnesses, near-misses), quality (deficiencies, non-conformances), maintenance (defects, rework) and human resources (absenteeism, turnover, grievances).

Loss prevention across these functions may be supported by a database of losses, incidents or undesirable events that enables investigation, identification of corrective and preventative actions, and tracking of these actions to close-out. Many off-the-shelf databases can be used to generate reports to inform managers of performance. Data may be dissected, for instance, by functions, operation, region or other criteria. A single online platform may be used to not only capture this lagging information (post-loss/event) but also schedule proactive monitoring processes. It's particularly important to provide assurance to the executive, boards and company officers, who are increasingly subject to scrutiny when companies incur losses that become public knowledge. The news regularly covers such events, for instance, breaches of statutory requirements resulting from workplace fatalities, environmental incidents that cause significant harm, or product quality failures that affect consumer safety.

Accreditation to one or more ISO Standards is one means by which organisations can improve their performance including in loss prevention. ISO was founded on a fundamental question about the best way of doing something. Some of the most commonly applied standards in industry include ISO 31000 – Risk Management; ISO 9001 – Quality Management; ISO 14001 – Environmental Management; and ISO 45001 – Occupational Health and Safety.

The latest versions of these standards require the organisation to define its context. The Channel assists with this by defining external forces generically. The greater the complexity of the environment of the enterprise, the more compelling the need to define context which often involves stakeholder analysis. These can include regulators, local communities, suppliers, contractors, strategic alliance partners and others impacted by the operations of the enterprise. Labour markets, union bodies and outsourcing suppliers may also be important.

Companies should recognise their own workforce as stakeholders. The analysis entails not only identifying these interested parties but also their needs along with the organisation's obligations to them, from a risk (and opportunity) perspective. For instance, "ISO 9001:2015 Quality management system – requirements" (International Organization for Standardization, 2015) contains Section 4 "Context of the organization" which covers determining external and internal matters that affect purpose, strategy and results; understanding the needs and expectations of stakeholders (who are referred to as 'interested parties'); deciding on the scope of the quality management system (QMS) and how the above two points will be considered; and establishing, implementing, maintaining and continually improving the QMS.

The above points are monitored and reviewed to assess performance and pursue continual improvement. To achieve accreditation to the standard, the organisation must have documented evidence of these processes and the associated outputs, for example a Stakeholder Needs Analysis. The business' QMS is reviewed by an external auditor and if the mandatory requirements are met to an acceptable level, accreditation is awarded. The enterprise is allowed to display their certificate, for instance, on its website so that clients and other stakeholders are made aware of this achievement. Some organisations see this as part of their competitive advantage, or at least a differentiator from unaccredited competitors.

ISO Standards set the expectation that managers will understand their organisation beyond the footprint of their operations. This external focus is aligned concurrently to internal management, resourcing and control requirements. The International Organization for Standardization explains the purpose and importance of each of its standards on its website, for instance: ISO 9001:2015- Quality management systems — Requirements; ISO 45001:2018- Occupational health and safety management systems – Requirements with guidance for use; and ISO 14001:2015- Environmental management systems – Requirements with guidance for use. The growth of ISO accreditation across various industries, countries, and regions is evident on the internet.

A practical caution, however, is needed. Just because an organisation has accreditation to a standard, that's not a guarantee against incidents or undesirable events (losses) related to that standard. It means that they have processes in place to manage the risk of those losses occurring, to recover and return to normal operations, to investigate and act for prevention, and to pursue continual improvement. The standards don't describe how the organisation will establish these processes only that they must have something in place. The efficacy of those processes is a matter for the organisation's leaders to address, not necessarily for the auditor to advise on.

AN ALTERNATIVE TO ACCREDITATION

Using the Channel as a self-assessment tool provides the organisation an alternative to the significant commitment associated with accreditation to external standards. It can enable the organisation to better understand itself and to prepare for external changes. Over the last 20 years, there have been transformational changes that impact the local level of the enterprise. The internet has provided workers and the public with greater access to information, and this has considerably influenced many of society's goals and values. The original Channel from *PSM1* included the two-way arrow at the bottom of the diagram between employees' goals and values and external environment forces. The interdependencies and feedback process between these two sources has strengthened, reinforced by the powers of social media. In previous years, if a fatality occurred at a workplace, it may eventually make a page in the newspaper. Now, it may be broadcast via social media within the hour. This has bolstered the necessity for risk management for many organisations, including SMEs. It also highlights the need to align workers' and the organisation's goals and values.

Today's managers should understand their organisation's risk exposure holistically, both as a technical system and social system. The singular concept of alignment doesn't work for operations because of organisational factors with sociological roots. Issues arise regularly at the operational level that are evidence of conflict over objectives and targets. A supervisor may want to keep a truck with worn tyres operating until the end of the shift to achieve more production, but the driver might not be comfortable with this decision for safety reasons. There are many examples which result from the inherent tension between what is safe enough and what isn't. Often the difference is due to perception rather than a clear understanding of energies and the adequacy of controls.

Overt and covert conflicts of this nature have a flow-on effect to the organisational culture. The same applies when companies ask for employee input but fail to follow-up on requests or to provide a reasonable explanation for why something can't be implemented. Alignment can fail when there's no or an inadequate feedback loop.

Even if OSH initiatives have the support of senior management, implementation can stumble at the lower management decision-making level. This is where a risk-based approach is most needed, not just for decision-making but also for risk-based communication.

(Solutions for these non-alignment challenges will be provided in Chapters 8 and 9.)

Conflicts at the operational level can lead to industrial disputes and legal action. This too is on the rise as society shifts increasingly to a litigation mindset and individuals are more aware of their rights. *PSM1* included the case of *Collins and AMWU v. Rexam Australia Pty Ltd.* (Industrial Relations Commission of Australia, 1996) as an example. Collins was a diabetic, which his employer knew prior to his employment. He had been hired on a 7:00 a.m. to 3:30 p.m. shift and he was able to manage his treatment regime around these times. When the company changed the shift in 1992/1993 to 9:00 a.m. to 5:30 p.m., Collins was asked to work this roster along with other employees. He declined to do so because it wouldn't suit his treatment programme. The company accepted this at the time after receiving a letter from the employee's doctor.

In 1995, the company changed its position and required Collins to work the new shift system. Following numerous warnings and correspondence because of his refusal, Collins' employment was terminated. The court found that the business had acted unfairly in dismissing him and had contravened S170DE (1) of the Industrial Relations Act 1988 (Australia) and awarded him with damages accordingly. The case showed that employers can't assume that employees will accept company practices where their needs are incompatible with them. Examples such as this, have increasingly become an organisational issue. Employers have OSH duties for both the physical and psychological health of workers because of recent legislative changes in some jurisdictions.

Few managers have the courage to undertake root cause analysis into decision-making processes to uncover underlying deficiencies within the management system, culture and/or leadership that could affect workers' psychological health.

Matters such as these have become more significant risks to businesses, including SMEs, due to the growth in access to knowledge. This has been accompanied by awareness that knowledge is power. In essence, the needs and expectations of the workforce have matured well beyond the employment/engagement contract. The psychological contract, whilst emerging in the early 1960s, has become paramount in the field of human resource management. This 'contract' refers to 'individuals' expectations, beliefs, ambitions and obligations, as perceived by the employee and the worker' (Chartered Institute of Personnel and Development, 2023). It's mostly been studied from the perspective of the employee, and it matters because it heavily influences day-to-day behaviour of employees. Whilst it's predominantly an area of focus in HR management, the overlaps with production, OSH and quality are evident in the advice of the Chartered Institute of Personnel and Development (CIPD):

Managers need to remember:

- Employment relationships may break down despite management's best efforts. Nevertheless, it's the managers' job to take responsibility for building and maintaining them.
- Preventing breach in the first place is better than trying to repair the damage afterwards. Trust is key to the relationship.
- Where breach cannot be avoided, it may be better to spend time negotiating or renegotiating the deal, rather than focusing too much on delivery.
- Interventions aimed at building resilience skills can help individuals cope better with psychological contract breaches.

(Chartered Institute of Personnel and Development, 2023)

The messaging clearly points to the critical role of the line manager in understanding and managing expectations of fairness. Where does this leave the supervisor when they override a worker's concerns about a safety or maintenance issue, such as worn tyres on the truck, to continue production for the remainder of the shift? The need for internal goal alignment, as shown in the right-hand side of the Channel, is greater now than when the Channel was developed 20 years ago. The question is how to make this happen. A decision-making tool is needed to consistently ensure that safety and production goals are balanced using a risk-based approach. The 'Reasonableness Test' (from *PSM1*) will be included in Chapter 8 of this second edition. Application of the test will assist line managers to justify operational decisions and set a standard for safety performance and production targets concurrently.

Risk-based decision-making must be at the core of transformation of management systems from silos to full integration.

For this to happen:

Traditional paradigms of 'safety' need to be challenged. The mindset change from 'safety' to 'risk management' is already underway and gaining momentum with the recognition that a multi-disciplinary approach to risk management is required. The opportunity is to apply this holistically to achieve production/productivity, work quality and HSE incident mitigation concurrently. A significant value-add is to continue the organizational and professional maturity journey with the people the EHS management system is designed to protect in the first place – frontline supervisors and workers. The challenge for business is how to tap into the depth of knowledge that exists at all levels of the organizational hierarchy and to gather the data to make well-informed risk-based decisions at both strategic and operational levels.

(Van der Stap, 2018)

LEARNING FROM THE ESG AGENDA

The discussions so far have highlighted weaknesses in current management systems that hinder production, safety and quality performance being pursued concurrently. Whilst described as 'gaps', these can be reframed as 'opportunities for improvement' or simply, progress in the right direction. The practical tools to sustain this transition towards a risk-based approach will continue to be presented in this book. These are:

- The Entropy Model explains the nature of risk and how residual and entropic risks affect multiple business functions, especially at the operational level. The model is a tool to manage risk more effectively, to drive a multidisciplinary strategic approach to risk management and to explain the importance of everyone's behaviours;
- The Alignment Fallacy explained that OSH policy doesn't cascade to 'Safety First' as a consistent, operational reality because of disruptions and conflicts down the hierarchy resulting from decisions that don't have a clear, consistent risk-basis;
- The Strategic Alignment Channel illustrated the need to align the enterprise with its external environment, which is changing dynamically and requires adaptability. Internally, resources must be balanced to achieve internal strategic alignment. The key consideration to attain internal goal alignment is creating a fit between enterprise and employee goals and values to sustain the psychological contract. There's a feedback loop to and from the external environment and the workforce suggesting that the organisation must be flexible to social changes and not focus singularly on being a technical entity.

Enterprise executives and senior managers need to be cognisant of this feedback loop along with the governing and influencing factors in the external environment. A significant growth area currently underway is the inclusion of the ESG Agenda in corporate reporting. The internet can be searched for an explanation of ESG including the history of development, the drivers behind the movement and how it's actualised in business. Much of the impetus has come from investors who have swayed towards corporations that take ESG seriously. Like the other matters discussed so far, ESG has a risk and opportunity basis that affects all functional areas of an enterprise, particularly those connected to the value chain.

The SME owner or manager may wonder what ESG has to do with them and the operation of their business. The answer is that such socially and commercially driven changes have momentum. Continual improvement is a clear expectation that will affect future supply chains. Earlier, the example was given of small businesses being required to sign a declaration binding to and assuring that they are compliant to the client's modern slavery policy.

High-level ESG commitments made by corporations will likely cascade to their suppliers, regardless of those suppliers' size or ability to meet expectations. The issue isn't can or will suppliers comply.
It's will they be ready.

For instance, how will an SME demonstrate X percentage reduction in greenhouse gas emissions? How will they develop and implement programmes for the mental health of their workers (currently enshrined in law in some countries)? How will they demonstrate best practice in hiring and onboarding to achieve safe, efficient, environmentally responsible production? It's not that these outcomes can't be achieved. The issue is readiness. Right now, functional silos and gaps in risk and opportunity management are holding organisations back.

THE MULTIDISCIPLINARY APPROACH

One of the major learnings from the COVID-19 pandemic was that no single business function had all the answers to the risks it presented. The impact was felt through the whole cycle of inputs, processes, and outputs from supply chain to internal operations through to the customer/client. The opportunity since the COVID-19 pandemic's major impact is to reframe what was learned from a risk management perspective and put a new lens on operational practices.

Another outcome of COVID-19 was that governments, corporations, SME owners and managers, and professionals with all different backgrounds that were supposed to have a sound risk basis (e.g. OSH, logistics and procurement), learned how much they didn't know. Most organisations would have been described as having a 'knowing culture' – "We know our business. We know what we're doing. We know our value". To the contrary, most organisations were thrown onto a sharp learning curve. It would be unwise to put this experience into the CEO's bottom drawer on the assumption that it won't happen again. In fact, it might be interesting to know how many organisations ran lessons learned workshops after COVID-19 to take away actions that better prepare them for the future. Opportunities such as these shouldn't be lost.

Commentary about this transformation was already taking place prior to COVID-19. According to McKinsey (2016), the challenge with moving from a knowing culture that relies on simple, efficient rules for making decisions (heuristics) to a learning culture isn't about the cost. It's about being more objective and data-driven. This requires imagination and inertia. The implication is that something needs to generate this approach.

The emergence of Resilience Engineering pointed to needed change:

> The term Resilience Engineering is used to represent a new way of thinking about safety. Whereas conventional risk management approaches are based on hindsight and emphasise error tabulation and calculation of failure probabilities, Resilience Engineering looks for ways to enhance the ability at all levels of organizations to create processes that are robust yet flexible, to monitor and revise risk models, and to use resources proactively in the face of disruptions or ongoing production and economic pressures.
>
> *(Resilience Engineering Association, 2017) (Olewoyin & Hill, 2018)*

As the discussion continues in the following chapters, the need for a different approach to risk management which is more holistic, pragmatic, opportunistic and that involves multiple disciplines will emerge.

SUMMARY

This chapter has considered the organisation from a sociological perspective in the context of the dynamics which occur in the external environment as well as internally. The 'Alignment Fallacy' was explained to provide understanding of why safety policy doesn't readily cascade to an operational reality, especially when such policy is founded on safety branding such as 'Safety First' or 'Zero Harm'.

One of the key issues associated with non-alignment of policy with how the work is done is the dilemma faced by frontline managers and supervisors who are responsible for safety performance and production targets in assigned work areas. To add to this, there are gaps in how supervisors develop relationships with new incumbent workers which involve inadequate consideration of the entropic risks (degradation) that occurs during onboarding and that dents the psychological contract. (This issue will be discussed further in Chapter 8 where the focus will be on the development of a risk-based culture and risk leadership competencies.)

In *PSM1*, a second tool accompanied the Entropy Model. This was the 'Strategic Alignment Channel' – a tool to assist business managers to better understand the context of their organisation in relation to the external environment, resourcing decisions and alignment with employees' goals and values. The Channel was designed to assist with management of change because it instils an awareness of governing and influencing factors beyond the operational footprint. The Channel is also helpful in analysing an enterprise as having inputs, processes, and outputs, as well as important internal and external relationships. This directs risk assessment more effectively with multiple functions (production, OSH, quality, procurement etc) focused on responding to major changes or events affecting the enterprise. Separate discipline-based risk assessments aren't effective in times of significant changes or threats.

Attaining ISO accreditation is another means of adopting this outward view of the organisation. One of the key requirements is to define the context of the enterprise and identify important stakeholders, their needs and the organisation's responsibility to them. Using the Channel is an alternative for pursuing improvements, especially for SMEs or larger organisations with scarce resources. With this outward and forward-looking mindset, the discussion segued to the emergence of the ESG Agenda as a growing trend that will cascade to organisations via changes in government policy, the behaviours of corporations, and community expectations. The key message was that a risk and opportunity approach is required to prepare for changes that are likely to impact all organisations in the Western and Developing Worlds. During COVID-19, businesses found that having a siloed, function-based response was limiting and instead, a multidisciplinary approach to risk and opportunity management was required to survive the pandemic. Current trends, such as climate change, the gig economy, technological innovations, including digital highways and artificial intelligence, health crises and other current and emerging issues, are driving a new paradigm of business.

A 'perfect storm' of developments increasingly requires that companies articulate and operationalize a corporate purpose that accounts for the impacts of their business on society and the environment.

(The Enacting Purpose Initiative, 2023)

This chapter has set the foundations for the enterprise being a sociological, not simply technological entity. This will be particularly important for Chapters 8–10 which discuss leadership, organisational capacity development, and the opportunities presented by the principles of high-reliability organisations, respectively. The discussions aren't intended to create grand strategies for organisations to move towards the lofty standards of the world's best corporations. It's more about optimising resources, both technical and behavioural, of enterprises to achieve much more from what they have – a pragmatic approach. A key theme throughout Chapters 1 and 2 has been operational reality and the need to displace delusional concepts such as zero harm, zero risk, the perfect leadership style and management systems that look fantastic on paper.

The following chapters will become more technically focused and explore in greater detail, the residual and entropic risks associated with systems of production. This will be done in Chapters 3–6. In Chapter 7, an alternative semi-quantification risk assessment method will be proposed to better understand the complexity, dynamism and numerous interfaces of risk as these occur in the workplace.

REFERENCES

Chartered Institute of Personnel and Development, 2023, Psychological Contract, Psychological Contract | CIPD.

EnactingPurpose.Org, 2023, epi-report-final.pdf (enactingpurpose.org).

Government of Western Australia, 2022, WALW - Work Health and Safety (General) Regulations 2022- Home Page (legislation.wa.gov.au).

Grieve, R., & Van der Stap, T., August 2019, Safety and entropy – a leadership issue. *Professional Safety Journal*, American Society of Safety Professionals.

Hasle, P, Madsen, C.U., & Hanse, D., 2021, Integrating operations management and occupational health and safety: a necessary part of safety science! *ScienceDirect Safety Science*, 139, 105247.

Hollnagel, E., 2009, The ETTO principle: efficiency-thoroughness trade-off: why things that (routledge.com).

International Labour Organisation, 2022, ILO declaration on fundamental principles and rights at work (DECLARATION).

International Organization for Standardization, 2024:

ISO - International Organization for Standardization

ISO - ISO 22000 — Food safety management.

ISO 9001:2015- Quality management systems — requirements.

ISO 45001:2018- Occupational health and safety management systems — requirements with guidance for use.

ISO 14001:2015- Environmental management systems — requirements with guidance for use.

ISO 31000:2018- Risk management — guidelines.

Industrial Relations Commission of Australia, Collins and AMWU v Rexam Australia Pty Ltd (960511) DECISION NO:511/96, available on Collins and AMWU v Rexam Australia Pty Ltd [1996] IRCA 533 (24 October 1996) (austlii.edu.au).

McKinsey & Company. 2016. How companies are using big data and analytics. https://www.mckinsey.com/business-functions/mckinsey-analytics/our-insights/how-companies-are-using-big-data-and-analytics

Olewoyin, R., & Hill, D., 2018, *Safety Leadership and Professional Development*. Park Ridge, IL: Society of Safety Professionals.

Rasmussen, J., 1997, Risk management in a dynamic society: a modelling problem. *Safety Science* 27(2):183–213. https://doi.org/10.1016/S0925-7535(97)00052-0

Resilience Engineering Association, 2017, About resilience engineering. https://www.resilience-engineering-association.org/

Tate, W., January 2013, Managing leadership from a systemic perspective, London Metropolitan University Business School ISSUE DATED 09 (systemicleadershipinstitute.org).

Van der Stap, T. 2018, Chapter 29 Risk leadership – a multidisciplinary approach. In R. Olewoyin, & D. Hill, *Safety Leadership and Professional Development*. USA: American Society of Safety Professionals.

World Economic Forum, 2022, 6 ways climate change is already affecting our lives. World Economic Forum (weforum.org).

3 Processes

THE INTERFACE BETWEEN SYSTEMS OF PRODUCTION

Work activities involve the interaction of human resources, technologies and the physical environment through processes. Processes are the interface between other systems of production. In practice, these can be divided into four main categories which are: systems processes; operational process designs or work practices; risk management practices; and post-incident practices.

The high-level processes that are usually proceduralised are those explaining how the system works within a functional area. For instance, OSH management systems tend to have overarching procedures for risk management, communication of risk, auditing, monitoring and inspection regimes, performance management etc. In Chapter 2, ISO accreditation was discussed. Historically, ISO Standards referred to having a procedure for selected mandatory elements such as internal auditing. The more recent versions refer to having a 'process'. Unsurprisingly, the easiest way to demonstrate having a process is to develop and implement a procedure, which is what the auditor will likely use to assess internal compliance.

The area of current contention in the OSH profession is the bureaucratisation of safety. The key issue is the disconnection between procedures written according to 'Work as Imagined' by people such as OSH professionals who don't perform the work, and 'Work as Done' by the workers who do the work and have practical knowledge of the required tasks and conditions they work in. Their input isn't always sought when preparing OSH documentation. Often OSH procedures reside separately from operational procedures even though there are strong synergies for combined documents. There are numerous academics who are championing this issue, so this debate is intentionally left outside the scope of this book. The focus here is primarily on operational improvements in risk management supported by top management. It's sufficient to say that whatever work is done should be managed on a risk basis.

OSH documentation should be developed and implemented to build operational capability towards achieving consistently safe, efficient production to expected quality standards, not for the glorification of the OSH function.

What's more important than the debate is building an understanding of process-related risks. It's worth considering how the Entropy Model might better inform the risks associated with a task. Key inputs include:

- The condition and characteristics (e.g. safety in design) of equipment used;
- The state and characteristics of the environment (e.g. safety in design) where the work is undertaken;

DOI: 10.1201/9781032701561-4

- The competencies and condition of individuals who do the work;
- The characteristics of the job/s or process/es undertaken;
- Any residual risks within these systems of production; and
- The interaction of these systems of production in terms of compatibility or incompatibility and the potential for loss of control.

The characteristics of the job or process warrant particular attention in this chapter. The primary concerns are that residual risks are managed, and entropic risks are prevented. The sudden degradation (and rising entropic risk) of one or more systems could result in losses, for instance, equipment failure, human error due to fatigue, damage to infrastructure etc.

Control of the process comes down to two things.

The first is the quality of the systems of production that will be used.

Second, how well the work has been planned, communicated, understood by those involved plus agreed contingencies if something changes during the work that could lead to losses.

As the discussion unfolds in this and the next three chapters which focus on each of the systems of production, the quality and reliability of these systems is shown to be central in achieving safe, efficient production to quality standards. It's because work is dynamic and ever-changing that it's challenging to consistently sustain the process system of production. New or modified work practices have a negative effect on safety and efficiency in the short-term because workers need time to adjust and learn new skills. In addition, some of the consequences of change might not be fully anticipated causing unexpected rises in entropic risk. This is where the inputs-process-outputs method is useful. For instance, a company may set up a quick and convenient vehicle wash-down area on a concrete slab at their facilities without considering the impact of hydrocarbon run off. This would introduce another source of entropic risk caused by the lack of fit between the process and the physical environment including other areas of the yard, a neighbouring property, into drains and eventually waterways. Following a process from start to finish helps to avoid poor decisions. For example, the following should be considered: the water source and the chemicals to be used for the washdown, the tools and equipment including any risks of using these (high pressure water; electrical equipment; chemicals), and the responsible disposal of wastewater.

Changes in the supply chain are also crucial, particularly, the reductions in the quality of inputs. For instance, a facility had an onsite sewage treatment plant. The procurement department decided to cut costs and purchased a cheaper toilet paper thereby saving the company $10,000 per annum. It was found that the paper didn't break down like the higher-quality paper. As a result, the entire system became blocked which cost the business hundreds of thousands of dollars to rectify. Input-process-output analysis with the involvement of multiple disciplines would most likely have prevented this costly mistake. Hopefully, the organisation's senior management didn't blame the procurement manager or view the incident as a one-off event. This simple example highlights the benefits of risk assessment when changes are proposed.

All processes have some level of human input and human fallibility is an accepted fact. At the operational level, the greater the direct involvement, the greater the worker's vulnerability to the risks associated with the interface between systems of production. In current practice, this likelihood takes exposure into consideration.

Exposure is often understood as a time-related variable but realistically, it's the severity of the risks more so than simply the length of time of contact that matters.

This is where the saying, 'being in the wrong place at the wrong time' comes into effect. The Entropy Model provides a better understanding of 'exposure' and relates it to a specific moment in time at which a process or task is being undertaken. What is more important than duration is the level of residual and entropic risks at that moment. Figure 3.1 illustrates this at various points in the model.

Three different moments are shown. Consider these as unrelated and representing three independent scenarios. At point 1, assume a high level of residual risk in the physical environment such as unstable ground conditions. Companies that have this type of hazardous condition can gather data through drilling but never have complete information about fissures, cracks or variability in rock hardness. The actual total residual risk contains some unknowns. Would a worker need to be exposure for a long period of time for the ground to collapse underneath them or could it happen instantly? Exposure time may be irrelevant in high residual risk situations. As a further example, in underground hard rock mining, there can be risks of toxic gas pockets, unstable sheets of rock on the workings roof, uneven ground, changing

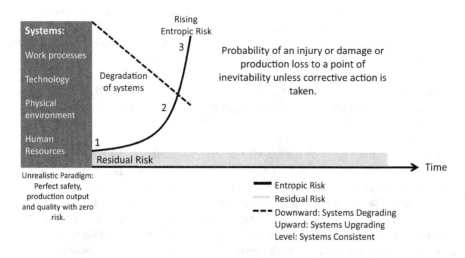

FIGURE 3.1 The second section of the Entropy Model with exposure points

water levels and other risks. The workplace is inherently dangerous and potential for an incident is significant and not necessarily time dependent.

At point 2, assume a moderate level of total residual risk in all systems of production. For instance, a driver is doing deliveries in a company vehicle. The driver has been at work since 7 a.m. and it's now after lunch, a time when mild fatigue sets in. The vehicle brakes are spongy, and the steering is loose, but the driver is used to this. It starts to rain lightly. Unbeknown to the driver, an accident occurred on the road the day before and 20 litres of diesel is still on the surface. All systems of production (at 2 in the figure) are in a moderately degraded state. Will an accident happen or not? Does the duration of exposure matter or is it simply that the combination of entropic risks intersects at a specific time in a dynamic process?

At point 3, assume the residual risk of all systems is moderate. One of systems of production is significantly degraded, for instance, in the driving scenario above either: the driver is intoxicated; or the vehicle brakes are starting to fail; or a large washout has occurred across the road. Exposure in terms of duration, in this scenario, is complicated. Being intoxicated doesn't necessarily mean that the driver will have an accident. Failing brakes don't mean the journey will end with an accident if a driver knows how to use the gears to reduce speed, responds by going up an incline and applies the park brake to bring the vehicle to a stop. The messaging here is that:

The probability of an incident relates to the absolute levels of residual and entropic risks at a given point in time, within a dynamic process. For this reason, a written risk assessment becomes redundant as soon as the work commences. Safety then depends on human responses to the changing risk profile. Risk variables must remain within tolerances that people can manage.

A worker when exposed to risk may be injured in the first or last five minutes or at any time in a day's work. The worker's susceptibility to injury is therefore related to the level of residual risk plus the variability in the entropic risk in other systems of production. The individual worker also has risks specifically related to their competencies. Areas in which they have a gap in knowledge, skills or abilities (KSAs) are a source of residual risk (due to the lead time for learning), whilst fluctuations in capacity are a worker-specific entropic risk.

Worker exposure is the sum of a complex set of variables associated with the systems of production.

The following equation provides a more realistic profile of worker exposure than using a risk matrix with consequences versus likelihood. This takes into account the residual and entropic risks of each of the four systems of production, including the risk levels of co-workers. This equation will be revisited at the end of Chapter 7 which presents a detailed alternative to the current risk matrix approach.

Worker Exposure at a given point in time =
Level of residual risk in the process, physical environment and technologies

+ Levels of entropic risk in process, physical environment and technologies

+ Level of competencies and physical ability of co-workers (i.e. others' residual risks)

+ Level of co-worker degradation (i.e. others' entropic risks)

+ Level of worker-specific competencies and physical ability (i.e. their own residual risk)

+ Level of worker-specific degradation (i.e. their own level of entropic risk).

(adapted from Van der Stap, 2008)

This process brings the systems of production together and creates the interfaces between these systems, for instance, at the person-machine interface. A poor fit at this junction creates demands on the person that may increase their rate of degradation, such as when complex control panels are more mentally demanding to operate than simple, well-designed panels.

This approach to worker exposure also takes into consideration the residual and entropic risks associated with other workers. For instance, an individual worker may be highly skilled and physically suited to the job (therefore a low residual risk). They may maintain themselves in a healthy, alert condition (therefore a low level of entropic risk). Their contribution to the total risk level within the workplace may therefore be minimal if their standards of human performance are maintained. If however, that individual in made to operate in a workplace with systems of production that have high residual risks and/or rising entropic risks from the physical workplace, technologies, high-energy uncontrolled processes, and/or other workers who are high risk, then that individual's exposure is high. The other worker/s may be considered high risk if they are inexperienced (high residual risk) and/or contribute high entropic risk due to states of degradation (such as drugs, alcohol, fatigue, ill-health etc.).

The complexity of the risk variables that contribute to a process means that when there is an incident, blaming the worker is not only unjust and foolish but also technically wrong.

The various inputs to a process or task can result in work being done well, as it often is – without incident or noticeable loss. There may, however, be other circumstances where the accumulative impact of risks whilst undertaking a process, exceed the capacity of the system to control or tolerate the risk thereby leading to loss. This may be evident as a safety incident, production loss, quality deficiency or other diminished return for the enterprise.

As mentioned earlier, the other types of processes that tend to be documented are risk management procedures and post-incident investigation procedures. Considering Figure 3.2, an investigation should include questions about the time prior to and at the moment of the loss event. The key areas are:

1. The residual risks inherent in the physical work environment, technologies and process, and whether these were managed effectively;

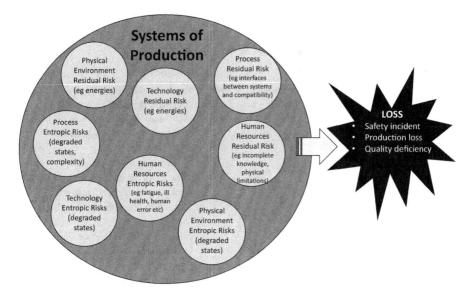

FIGURE 3.2   Systems of production and the complex risks contributing to losses

2. Any degraded states in the physical environment, technologies and the process, and the severity of this degradation;
3. The residual risks inherent in the workers undertaking the job and whether these were managed effectively;
4. Any degraded states in the workers and the severity of this degradation.

Traditional approaches to incident investigation tend to start with the worker who was directly involved in the incident. This new approach involves understanding context in which the undesirable event occurs before questioning any factors related to the person.

Figure 3.2 shows the systems of production and the residual and entropic risks of each, resulting in eight risk sources that can contribute to losses. Within the total system, human error is one element within HR entropic risk. Incomplete KSAs are a key component of HR residual risk. From a statutory perspective, the employer is accountable for ensuring that workers have the competencies they require to work safely. The employer also has a duty to ensure personnel aren't suffering from fatigue or other form of temporary incapacitation. The figure infers, given the relative insignificance of intentional human deviance within the entire system, that blaming the worker isn't technically correct when undertaking an incident investigation, as there are many other potential contributing factors. (This will be discussed further in Chapter 6 on HR risks and safety violations.)

CASE STUDIES: DEEPWATER HORIZON AND TRAIN DERAILMENT

Table 3.1 provides an example of an incident and how the above factors may have contributed to the loss. The case study used is the Deepwater Horizon Disaster that

TABLE 3.1

Deepwater Horizon Categorisation of Risks

System of Production	Residual Risks	Entropic Risks
Processes	High residual risk associated with the main process which involved extraction of hydrocarbons under pressure from under the seabed.	Investigation findings included the following degraded processes: "weaknesses in cement design and testing, quality assurance and risk assessment" "proper integrity testing of the shoe track, production casing and casing hanger seal assembly were not undertaken" "If fluids had been diverted overboard, rather than to the MGS, there may have been more time to respond, and the consequences of the accident may have been reduced" "potential weaknesses in the testing regime and maintenance management system for the BOP"
Technologies	The main production technologies involved pressurised systems. One of the major residual risks on rigs is the potential for hydrocarbons including gases to contact ignition source/s. "Hydrocarbons migrated beyond areas on Deepwater Horizon that were electrically classified to areas where the potential for ignition was higher. The heating, ventilation and air conditioning system probably transferred a gas-rich mixture into the engine rooms, causing at least one engine to overspeed, creating a potential source of ignition"	Investigation findings included the following degraded technologies: "annulus cement probably experienced nitrogen breakout and migration, allowing hydrocarbons to enter the wellbore annulus" "potential failure modes that could explain how the shoe track cement and the float collar allowed hydrocarbon ingress into the production casing" "The BOP emergency mode did not seal the well. Three methods for operating the BOP in the emergency mode were unsuccessful in sealing the well."
Physical Environment	The rig was located in a high residual risk environment including subsea and oceanic conditions. This was also a factor in relation to emergency response and evacuation.	The physical work environment (the rig) went into a rapid state of degradation when: "hydrocarbons were vented directly onto the rig through the 12 in. goosenecked vent exiting the MGS, and other flow-lines also directed gas onto the rig. This increased the potential for the gas to reach an ignition source"
Human Resources	The competencies of the managers and work crews to prevent and deal with the emergency were not discussed in the report.	The condition of the managers and work crews were not discussed in the report.

occurred in the Gulf of Mexico in the evening of 20 April 2010. BP's Accident Investigation Report doesn't discuss the competencies or condition of the managers or workers, so the commentary here focuses on breaking down the residual and entropic risks associated with the processes undertaken, the physical work environment and the technologies in use prior to and during the disaster as it unfolded. For this reason, the list is only a sample of the actual event based on the report from BP available at: deepwater-horizon-accident-investigation-report-executive-summary. pdf (bp.com) (British Petroleum, 2010).

This example illustrates that the post-incident investigation can be broken down into the eight sources of risk (residual and entropic risks in each of the four systems of production). These combinations can also be used to better understand the risks that affect people at work. As was shown in Figure 3.2, it contextualises human error as a HR entropic risk within the total production system.

Instead of starting an incident investigation with what the person/s did wrong at the time of the undesirable event, the question shifts firstly, to analysis of the system and circumstances that led to one or more critical risk decisions.

The second consideration is whether those person/s were adequately prepared (by the system) to manage or control the risk state that led to the event.

The table presents each system of production in the left-hand column. The next column is the associated residual risk with a description of evidence from the report. The table shows these were high in processes, technologies and the physical work environment. Specifically, high residual risk was associated with the main process which involved extraction of hydrocarbons under pressure from under the seabed. The main production technologies involved pressurised systems. One of the major residual risks on rigs was the potential for hydrocarbons including gases to contact ignition source/s. The rig was in a high residual risk environment including subsea and oceanic conditions. This was also a factor in relation to emergency response and evacuation.

The righthand column identifies the entropic risk of each system of production, again with quotes from the report that provide evidence. There were numerous degraded processes including weaknesses in cement design and testing, quality assurance and risk assessment. Integrity testing of critical components wasn't undertaken along with weaknesses in testing regimes and maintenance management systems for a key part of the plant. This resulted in degraded technologies. The rig (being the physical workplace) went into a rapid state of degradation when gases were released and reached an ignition source.

Given the risk states of these systems of production, the actions of the workers become proportionally insignificant. The fact was that the high residual risks and rising entropic risks had reached a point of inevitable disaster. It's doubtful that the workers alone could have done anything to influence this state.

The second case study is from a train derailment in Western Australia. This also provides context for human fallibility. Current 'duty of care' based legislation doesn't allow the employer to lay sole blame on the worker or workers. The following news story about a train derailment illustrates this point.

A fully laden ore train pulling 268 carriages was derailed on 5 November 2018. The driver was out of the cabin when the train started travelling more than 90 kilometres before it was intentionally derailed by the remote-control centre which was 1,500 kilometres away. The driver's employment was terminated by the company. An unfair dismissal case was lodged (ABC News, 10 January 2019, BHP driver sacked after Pilbara train derailment claims unfair dismissal – ABC News). An undisclosed settlement was reached six weeks later (ABC News, 2019, BHP reaches settlement with driver sacked over iron ore train derailment in Pilbara – ABC News).

The investigation by the Australian Transport Safety Bureau found that there were several issues. A rail maintenance team had earlier started applying brakes to the wrong train. The team had been sent to assist the driver because the train had a disconnected cable, but they started work on another train nearby. In the meantime, the faulty train started to roll away.

Had the team gone to the correct train, it would have still moved away because the emergency brake hadn't been applied properly in the front locomotive and there had also been an electronic failure. The driver made an emergency call and the train was derailed remotely (ABC News, 12 March 2019, Handbrakes applied to wrong BHP iron ore train before Pilbara runaway derailment, ATSB finds – ABC News).

The regulator identified gaps in risk assessment processes that meant ore trains were vulnerable to a runaway event because potential causes and critical controls were inadequate considered. Safety critical actions weren't clearly communicated to drivers.

The company reports having implemented a new rail system engineering framework with integration of train braking systems and operating protocols. Human factors related to fatigue and rosters were also considered (ABC News, 17 March 2022, ATSB train derailment report raises concerns about BHP risk assessments and communication – ABC News).

When considering the incident investigation process objectively, there are several questions to ask about boundaries of accountability between the organisation and the person/s involved:

1. Who has control over the systems of production?
2. Who is accountable for ensuring those systems of production are safe and operate in an environmentally responsible manner to required quality standards?
3. Who is accountable for ensuring the people engaged to perform work have the required competencies, i.e. the residual risk of individuals, teams and the workforce as a whole?
4. Who is accountable for ensuring that if there are competency gaps, there is adequate supervision and instruction in the short-term and training over the longer-term?

5. Who is accountability for ensuring that people can maintain the capacity to do such work safely and to required standards, i.e. the entropic risks of people are prevented sufficiently?
6. Who is accountable for monitoring people to ensure they aren't suffering from degradation?
7. Who is accountable for compliance once the company has taken all reasonable steps to make sure the workers are operationally prepared for the risks of the work?

The answer to questions 1–6 is the organisation. The answer to 7 is the individual worker. Each situation would be looked at by the regulator or the courts on a case-by-case basis. For this reason, on the assumption that the organisation has fulfilled its obligations towards the worker, the question following an incident should be:

How did the organisation fail to protect a competent, conscientious worker?

Organisations need to take more responsibility for the quality of their HR system of production; after all, they control all steps in the engagement process from recruitment and selection, through to training, on-the-job supervision, performance management etc. In this context, the enterprise should start from a baseline of objectivity when investigating the behaviours that preceded the undesirable event.

MINIMAL COMPLIANCE

A review of local OSH legislation is likely to reveal a balance between the accountabilities of the organisation, the worker and other duty holders. The Strategic Alignment Channel discussed in Chapter 2 indicated that enterprises are required to be legally compliant and socially responsible driven by governing and influencing forces in the external environment in which they operate. Some of the major transitions in OSH legislation are worth noting, the most significant of which was the shift away from prescription-based legislation to 'enabling' legislation. This shaped current approaches to the legislative framework in the UK and Australia, in particular. This incorporated the concepts of 'duty of care' and 'reasonably practicable', because of the Robens Report of the early 1970s.

> Robens went away from listing lots of individual regulations telling you exactly what you ought to do about particular hazards and the risks that arise and made it a simple principle – goal setting – every employer should identify the risks that they're generating and manage them effectively.

> *(British Safety Council, 2022)*

Managers need a sound knowledge of OSH laws that apply to their operations, and, where applicable, this includes accountabilities that apply to them as an 'officer'. In Australia, an 'officer' has the authority to make decisions that affect the whole, or a substantial part of a business or undertaking. Such persons have a duty to ensure appropriate systems of work and must actively monitor and evaluate work health and safety (WHS) management (WorkSafe Western Australia, 2021). They must exercise 'due diligence' in relation to WHS matters to make informed decisions when discharging their duty. Note the use of the term 'must' as opposed to 'should', meaning that 'must' is mandated by WHS legislation.

This legislation continues to evolve especially in response to major disasters such as the Grenfell Tower Fire. Psychosocial risks are now also legislated adding to the depth and breadth of the OSH Professional's duties to keep management informed and provide advice on risk mitigations. An underpinning principle is the concept of 'reasonably practicable'. In the UK, the term 'so far as is reasonably practicable' (SFAIRP) is used (Health and Safety Executive, 2024). 'Reasonably practicable' addresses:

- The severity of any potential injury or harm to health and degree of risk of this occurring; and
- Knowledge of the injury or harm to health; the risk of it occurring; and the ways to remove or mitigate the injury or harm; and
- The availability, suitability, and cost of removing or mitigating the risk.

The bottom line is that when it becomes apparent that there's a risk of injury or harm, the organisation's management must do something about it. If however, management should know about it, for instance, because it's industry or community knowledge, then not knowing or ignorance isn't an excuse for inaction or a leave pass from accountability. Where the Regulatory Authority finds a breach, prosecution action may be initiated. The penalties imposed through this process are separate from claims of negligence by a third party against the company. In these circumstances, a Court of Law will test whether the enterprise has been negligent using the 'reasonably foreseeable' and 'proximity' tests.

Enabling WHS legislation is performance-based because it allows and encourages organisations to actively develop their own workplace-specific safety standards. Enterprises committed to self-regulation and achieving higher levels of safety and social responsibility than required by external forces exhibit proactive strategies. These can include strict operational discipline in the areas of planned maintenance, monitoring regimes including internal audits, and continual improvement processes. The extent of self-regulation implies that little intervention is required from the regulator' however, as was shown by the Entropy Model, organisations can degrade again following external and/or internal changes, putting them at risk of losses that may draw regulatory attention.

THE PATH TO MATURITY AND SELF-REGULATION

PSM1 provided Figure 3.1 to illustrate a simple process management continuum and compliance outcomes. This is redesigned in Figure 3.3 to point to the end-goal of achieving an organisation which operates on a risk and opportunity basis rather than according to siloed functions that are potentially in conflict.

There are models available in OSH literature that classify organisations according to their maturity in relation to OSH management, compliance, and culture. OSH maturity models perpetuate the siloing of the function and ignore safety's place as one of many disciplines that belong under risk and opportunity management. A broader strategy should be adopted that is sensitive to operations and highly concerned with maintaining states of safety with reliability.

In this new order, organisations should be learning constantly and revering those enterprises that successfully adopt the principles of high-reliability organisations, employing a risk and opportunity principal strategy.

It would be expected that with this approach, documentation of 'The Work' would be fully integrated so that operational work is seamless with control of OSH, environmental and quality risks associated with that work. Risk and opportunity assessment processes should be concerned with complexity, uncertainty and potential high-risk environments. Organisations should focus on having sufficient capacity to navigate and respond effectively to changing demands.

At the operational coalface, it's impossible to separate safety from reliability. For instance, remote exploration involves the mobilisation of plant, support equipment, vehicles and crews to locations that are far from community infrastructure and services. If equipment breaks down because of poor maintenance, the risk of safety incidents and schedule delays escalates significantly. The pre-mobilisation inspection should be the last verification undertaken well ahead of leaving the controlled facilities of the operation's homebase. Unfortunately, due to schedule pressure (and underlying poor planning), this process may be rushed.

Defence operations are likely to have the same issues where lack of technological reliability leads to escalation of entropic risks putting people in danger and mission objectives at risk of failure. For this reason, defence operations have structured ways of considering all the factors associated with the task at hand – "the military Estimate of the Situation formed the cornerstone of an army officer's function" (Stogran, 2016).

Figure 3.3 takes into consideration the 'Safety Journey Maturity Model' (Hudson, 2003, 2007; Foster and Hoult, 2003). Hudson et al. mapped five steps later included by Van der Stap in Chapter 29: "Risk Leadership: A Multidisciplinary Approach" (Olewoyin & Hill, 2018):

Risk Tolerance	Hudson's Maturity Level	Operational Reliability within Systems of Production	Strategic Focus	Foundations
Low	Generative	High	Principal Strategy	Profit (Sustainable) Integrated Excellence
	Proactive		Aspirational Strategy	Profit (mid to long horizon) Functional Excellence
Moderate	Compliance	A result of compliance	Functional Strategy	Profit (medium horizon) Functional Performance
	Reactive		Survival Strategy	Profit (short horizon) Avoid External Flack
High	Pathological	Low	No Strategy	Profit

FIGURE 3.3 The path to the principal risk and opportunity strategy

1. at the lowest level is a 'pathological' stage where systems are under-developed and the culture is 'no care';
2. improving to 'reactive' only taking risk management seriously after incidents;
3. maturing to 'compliance' is mostly driven top-down;
4. becoming 'proactive' involves worker participation bottom-up; and
5. culminating in 'generative' where risk is managed as a way of organisational life.

The figure starts on the left-hand side with the risk tolerance of the enterprise ranging from low to moderate to high. These align to Hudson's Maturity Level with 'Generative' having a low risk tolerance, 'Compliance' having a moderate tolerance (only meeting basic legal requirements), and 'Pathological' being associated with high risk tolerance (non-compliant). In the centre column, these correlate to high and low reliability at the extreme ends of the spectrum. Unsurprisingly, reliability is driven by compliance at the moderate tolerance level. It's proposed in the figure that these also drive the strategic focus of the enterprise. This ranges at the bottom where there's no strategy, through to functional strategy for compliance. Progressively as the organisation becomes more proactive, the strategy becomes aspirational until a high reliability environment is created. Reliability and safety of the systems of production becomes central to the principal strategy. Timeframes shift accordingly. The pure profit motive drives enterprises at the bottom end of the spectrum. Sustainable profit and integrated excellence drives enterprises at the top end of this continuum.

There are underlying cultural factors that will influence how processes are determined. These have a flow on impact on operational risk and opportunity management. Figure 3.3 proposes a pathway for improvement. It's a vision for a dynamic, outwardly aware, organisational environment. Opportunities lay in building on previous work undertaken in business management and culture development across industries, and in particular, the emergence of high-reliability organisations as role models for enterprises aiming for sustainability.

How can this be achieved? Organisations should firstly understand where they are on the continuum then aim to progress incrementally to the next level up. For instance, the business operating reactively based on survival strategy should aim to mature beyond profit on a short horizon. Currently, their management may be motivated by avoiding external flack or accountability (being publicly found out) because of a lack of understanding of risk in the broader business content. This gap can be closed through a facilitated enterprise level risk assessment to better understand their organisation in a broader context, covering factors such as stakeholder analysis, supply chain risks and other strategic matters, that progressively cascade to operational challenges.

In the mid-level, the organisation with a moderate risk tolerance may be compliant where the law clearly dictates and 'fuzzy' in matters of self-regulation. Reliability is an outcome of compliance. For instance, legislation prescribes inspection of plant and equipment on a risk basis. The organisation will have paperwork as evidence of this; however, how well it's done will directly impact whether the plant and equipment are actually safe and reliable. At this level, the strategic focus is functional so operations achieve what they can whilst support functions such as OSH do the same with the resources provided. Management is focused on the enterprise being in business for the medium-term to the extent that it understands the environment that it operates in.

As the organisation matures, horizons are scanned for 'best practice' within their industry or through potential learnings across industry sectors. Profit timeframes stretch from mid-term to longer horizons. Each function is expected to pursue excellence. There's a desire for synergies across functional boundaries.

At the highest 'generative' level, operational reliability of all systems of production is a primary focus. Continuity of production, safety performance and quality assurance can't be achieved without it. Simply, lack of reliability is an indication of rising entropic risk and potentially unmanaged residual risk. The profit objective remains focused on sustainability and is the foundation of how the organisation develops its Principal Strategy. Sustainability infers an outward and inward focus concurrently based on alignment.

Where an organisation sits on this continuum will shape how it approaches the development and implementation of its processes to control risks (and exploit opportunities) at the operational level. Even simple processes can send powerful messages. For instance rudimentary practices such as whether work teams hold 'safety meetings' or 'safety and operations meetings', plus how often these are conducted – both planned and as needed, convey management's mindset. Psychologically, this basic practice can impact workers' perceptions regarding the compatibility of the two functions. If concurrently, they regularly observe operations managers and supervisors working closely with the OSH team in the workplace, this lends weight to OSH being integral to operational performance. In contrast however, if the OSH Advisor comes to the workplace on a weekly basis to conduct the 'safety meeting', safety will be seen as a separate silo and after the meeting, operations will resume business as usual.

The 'devil is in the detail' when it comes to messaging about safety and production at the operational level, so it's important that senior management clearly define their position in relation to compliance versus 'beyond compliance'.

The decision to only meet minimum standards is neither right nor wrong. It's based on risk, but the assurance processes must be in place for it to be a legitimate business strategy. Senior management would want to know that there's no slippage into sub-standard enterprise performance, however, the Entropy Model indicates that this can occur easily. A high jumper doesn't aim for the bar. They aim above the bar to make sure they don't commit a foul.

What may become problematic is where self-regulation drives industry standards. For instance, in some international mining companies, the standard for site vehicles until 2016 included modifications such as internal rollover protection, bull bar, flag, beacon, and reflective body stripe. Such companies required that all contractor vehicles going to their sites were equally fitted out so this became industry standard across most of the Australian mining industry. The cost was incurred by businesses large and small.

In 2016, one of these companies changed to Five Star ANCAP rated vehicles for their own fleet as well as contractor-owned or hired vehicles (Australasian Mine Safety Journal, 2013; Australian Mining, 2013). Inspection processes were established on company sites to ensure compliance to the light vehicle standard, with any failures resulting in access being denied. At the time, some commonly used vehicle brands were not Five Star rated leaving contractors in a financial quandary on how to retain service contracts. In 2023, ANCAP (the independent standards authority) changed the criteria for ratings. As a result, a vehicle sold in 2017 with a five-star rating may not achieve this level against the new criteria (The NRMA, 2023). This simple example highlights the complexities faced by businesses from external factors that are beyond their control and the cost implications of 'self-regulation'.

COMPLEXITY AND MANAGEMENT OF WORK PRACTICES

It's generally accepted that the external environment of the organisation is changing dynamically, and the level of complexity has risen significantly over the last 50 years. In the meantime, many of the processes undertaken at the work site remain fundamentally the same or similar in relation to fatality potential. This is except where there have been major changes in technologies that have removed the worker from hazardous situations altogether. For instance, on a construction site many activities remain consistent. These include excavations; overhead crane operations; erection of scaffolding; installation of services such as electricity; introduction of hazardous substances; and access/egress with traffic management issues.

Methods for the control of risks associated with these 'major hazards' or 'fatal risks' is embedded and readily available in industry. Fatalities, however, continue to occur and the question needs to be asked whether this is due to the complexity of processes or systems to manage the work. Sandone and Wong (2010) undertook an interesting case study of an electric utility organisation in Canada.

CASE STUDY: CANADIAN ELECTRIC UTILITY

The study involved using a 'Sense-Making Framework Approach' to classify safety interventions using four categories being 'Simple', 'Complicated', 'Complex' and 'Chaotic'. (These relate to the process system of production.) The organisation was experiencing changes due to external and internal forces. Despite, in the past, having a good safety record and culture, there had been an increase in accidents on site, which alarmingly, included fatalities.

According to the researchers, 'Simple' systems are easy to understand with a clear relationship between cause and effect. These are known knowns which are easy to comply with. In contrast, it's difficult to quickly comprehend the cause and effect of complicated systems. Experts might find these useful but operationally, because workers don't readily know the logic behind the experts' modelling, these may come down to a set of rules. 'Simple' and 'Complicated' systems are orderly. If this is considered in relation to the Entropy Model, workers can comply with simple instructions and follow rules, so therefore uncontrollable rising entropic risk shouldn't occur within the process system of production.

The Work becomes problematic however, when 'Complex' systems are introduced. It's impossible to understand the relationship between cause and effect, which can result in unpredictable behaviours. In the study, it was proposed that these responses couldn't be forecast but patterns could be observed. Though not stated, it's assumed that these may have been evident as near-misses or other losses. Finally, 'Chaotic' systems had no relationship between cause and effect. Both 'Complex' and 'Chaotic' systems were disorderly. Relating this to the Entropy Model suggests that these would also contribute to degradation of behaviours when undertaking processes and therefore, cause rising entropic risk.

The conclusions of Sandone and Wong's (2010) are particularly poignant in this discussion about control of activities at the operational level. They stated:

> Chaotic domain required fast action to prevent a catastrophe from happening. Complex domain issues require experimentation as a way to learn about and understand its patterns; patience is required to let the system find a practical and acceptable solution. Applying an improper decision-making approach can cause counterproductive effects. As the authors discovered, the two most recurrent safety issues that emerged from the analysis (safety support and safety rules) seem to be a direct consequence of the organization's strong willingness to reduce safety incidents by introducing more rules to increase compliance. This approach is a typical Ordered side solution and fails to consider the complexity of work situations. Safe-fail experimentations (for safety support) and consensual expertise (for safety rules) have been suggested as ways to eliminate ambiguous, confusing and frustrating work situations that lead to unsafe practices.

(Sandone and Wong, 2010)

The first point flags the need for corrective action when chaos (or rapid degradation) becomes evident. The second point about complex domain issues and 'experimentation', infer data gathering of patterns and recurrent issues, which according to the Entropy Model, would result in 'maintenance practices' to control complexity using a 'practical and acceptable solution'.

The final significant learning from this paper is that increasing the number of rules can cause complexity to rise, which may be counterproductive. Leveson (2011) has written at length about how new technologies have introduced complication and new causes of accidents are arising from it. This includes avoiding treating complexity as one indivisible property as there are various types when managing safety. In the field of human factors in safety performance, there has been extensive work done on 'Complexity management theory', 'Sensemaking' and the 'Implications of complexity in managing safety culture' (Carrillo, 2011).

A strong message that can be deduced from this case study and the other research work undertaken in the last 15 years in combination with the author's experience is:

Operations must be protected from chaos and complexity.

This comment raises questions about the role of risk-related professions who provide support services to operations, such as OSH, environment, quality and contractor management. Change in inevitable and in large organisations often comes from head office or senior management (top-down). Sometimes new strategies or programmes aren't explained well or contextualised or fully considered from an operational implementation perspective. Without the background detail, talking to operations about impending changes is likely to be met with uncertainty and apprehension about the additional work, potential chaos and/or complications that will be introduced. For this reason, operations should be protected from chaos and complexity, until well thought out solutions can be socialised.

As a fundamental principle of risk management, operations should be enabled to maintain control over critical risks that affect their work. Supporting functions should provide a buffer from influencing forces that are beyond operations' control.

Solutions should be socialised framing them from a practical perspective.

Much more could be written about the impact of complexity on risk, based on OSH and other literature. It's more meaningful and useful, however, to consider these matters within the everyday context of work, especially at the 'coalface' by asking basic questions:

- "Do any of our processes create chaos?"
- "Are our systems too complex to be implemented as intended?"
- "If yes, to either or both, what can we do about it, and what is the risk if we don't act?"

From OSH experience, the appetite for complexity can arise from incident investigations. This was confirmed in the case study, "*a direct consequence of the organization's strong willingness to reduce safety incidents by introducing more rules to increase compliance*" (Sandone and Wong, 2010). Operations managers and OSH professionals should be sceptical about the practical validity of ALARP when more rules and procedures are introduced. Unfortunately, ALARP is currently the accepted basis for risk reduction in industry.

THE ALARP ASSUMPTION

ALARP was defined in Chapter 2. As explained by the Health and Safety Executive UK, ALARP and SFAIRP (so far as is reasonably practicable) are essentially the same as at their core is the concept of 'reasonably practicable' (Health and Safety Executive, 2023). In the USA, ALARA – 'as low as reasonably achievable' – is used which was developed as a guiding principle for radiation safety (Centers for Disease Control and Prevention, 2024). A summary table of countries, organisations and their terminology is in the paper, ALARP-Final-Paper-Publishing.pdf (Mary Kay O'Connor Process Safety Center, 2020). The key findings were:

- A number of developed countries with mature process safety cultures are adopting, or have adopted, ALARP or similar principles as risk acceptance criteria.
- Although the underlying principles are similar, the application of the principles varies significantly.

- ALARP principles are often applied differently to various industries even within the same country.
- Both qualitative and quantitative approaches have been applied for risk estimation.
- All approaches recommend cost-benefit analysis, but there is no consensus on exactly how it should be done.
- There is no clear guideline on how to determine the disproportionality factor (in cost benefit analysis).
- Data is very scarce on how ALARP is practiced in industry.
- Although ALARP is not mandated in the United States, some US based companies practice ALARP principles, probably because of the global nature of the companies.

(Mary Kay O'Connor Process Safety Center, 2020)

As a global standard, risk continues to be viewed as a singular concept that relates from an OSH perspective to the nature of injury, the likelihood of injury and the period over which it occurs (CCPS, 2009). The implementation of ALARP is facilitated by using the Hierarchy of Controls (HoC), an explanation of which can be found online (Centers for Disease Control and Prevention, 2023 and NIOSH 2023).

The rationale of the HoC is that controls at the top of the triangle – elimination, substitution and engineering controls – are more effective and preferable to those at the lower end – administrative controls and personal protective equipment (PPE). In practice, what becomes problematic with the combination of the definitions of ALARP and 'reasonably practicable', with having a 'WHS duty', is that organisations know they must do something (not nothing), in relation to a risk of injury, but cost is also an accepted critical factor in the definition of 'reasonably practicable'. It's relatively easy therefore for organisations, following an incident investigation to do 'something' which is low cost. Concurrently, they can argue that a new level of ALARP has been achieved because of the implementation of inexpensive additional controls. What tends to happen is administrative controls and PPE changes become attractive options.

Unfortunately, in practice, these lower order controls are highly dependent on the worker following a procedure and wearing PPE to a standard of performance. It's widely accepted that human beings are inclined to introduce inconsistencies in how work is done. This same rationale applies to undertaking safe work practices. There will be variability from person to person in how they execute the task. According to the Entropy Model, this is because of competency deficiencies (residual risk) and/ or physical limitations (residual risk) and/or entropic risks such as fatigue, errors in judgement and other human factors. The quandary is if administration controls and PPE compliance are not reliable, how can a new lower level of ALARP be achieved by adding more of these controls?

This creates a dilemma resulting from the disconnection between principles or definitions and operational practice. This can be labelled the 'ALARP Assumption'. It's assumed that further layers of administrative controls reduce risk when it may have the oppose effect if the system is becoming overly complex to implement.

A measured approach is required when considering the 'ALARP Assumption'. Firstly, there are some administrative controls such as permit to work systems that are critical to manage residual risks such as containment of hazardous substances and isolation of energy sources. The value of these shouldn't be underestimated. It's OSH complacency and *"the organization's strong willingness to reduce safety incidents by introducing more rules to increase compliance"* (Sandone and Wong, 2010) that require a rethink in relation to actual, operational risk reduction. The gap is likely the failure to prove that safety and reliability within the systems of production have improved as a result of these additional lower order controls.

> *Complacency and over-zealousness may be opposites, but both might lack objectivity when it comes to understanding the impact of change on operational risk levels.*

It's not being suggested here that the global definition of ALARP or SFAIRP or ALARA needs to be overhauled from a regulatory perspective, however, in practice, the limitations need to be carefully considered particularly for operational risk owners. When incident investigations result in superficial causes and corrective actions that involve further procedural controls or rules, questions need to be asked:

- Has there been a reduction in actual risk because of these added controls; and
- Is there now a greater level of reliability in the system to prevent future losses or failures.

More broadly in relation to current operational practice:

- Do additional procedural controls reach a point of diminishing return in terms of risk reduction?
- Do they add complexity that reaches a point where the risk rises rather than falls because human capacity to cope is exceeded?
- Do they reduce efficiency by adding low value thoroughness to the system?

The latter question takes into consideration that an organisation working on a risk basis to pursue production, safety and quality performance concurrently should know when efficiency is being compromised for thoroughness because there is a trade-off, as per Hollnagel's ETTO principle (2009). From a risk perspective, minor reductions in safety risk due to more thorough operational risk assessment or processes, may generate time pressures that result in rising 'production pressure risk'. A risk transfer may occur rather than a risk reduction, thus defeating the aim to make the work safer.

The ALARP Assumption may also apply during the design and construction phases of a facility, well before it reaches the operational phase. For instance, the new major hazard facility project in the design stage usually involves complex hazard analysis methods that determine ALARP. These methods can be described as process safety:

A disciplined framework for managing the integrity of operating systems and processes handling hazardous substances by applying good design principles, engineering, and operating practices. It deals with the prevention and control of incidents that have the potential to release hazardous materials or energy. Such incidents can cause toxic effects, fire, or explosion and could ultimately result in serious injuries, property damage, lost production, and environmental impact.

(Center for Chemical Process Safety, 2009)

Process Safety focuses on engineering and design throughout the life cycle of this type of facility. There are numerous risk assessment processes undertaken and the most common are described in the glossary of the CCPS (American Institute of Chemical Engineers, 2024). One is the Hazard Identification Study (HAZID) which is based on the front-end engineering design (FEED) to enable progress to detailed design. The key objectives are:

- To identify inherent hazards early in the plant design development;
- Quantify the risk associated with each hazard;
- Based on the level of risk, initiate a process to ideally eliminate the hazard or reduce risk to a level to ALARP through elimination, substitution or engineering;
- Identify safety critical elements of the preliminary design;
- Commence the functional safety lifecycle to determine the extent and integrity of the plant's safety instrumented systems (SIS).

The HAZID is a critical piece of work for new projects because it informs the design strategy. For instance, at FEED, decisions need to be made that affect the cost of construction. This cost is weighed against the as-built risk profile and running costs in the operational phase. Operations may need to undertake modifications to plant or infrastructure, or work around design deficiencies, such as using scaffolding during shutdown maintenance. Ideally, safe working at heights would have been achieved by installing platforms, walkways, and ladders during construction. Whilst it's generally agreed that it's cheaper to build safety at the design and construction phases rather than retrofitting during operations, this doesn't always happen because a higher cost build makes it more difficult to attract investment capital. This creates a trade-off regarding 'reasonably practicable' and an ethical dilemma for design engineers.

In many jurisdictions, project regulatory approvals processes, both safety and environmental-related, mandate the completion of HAZID studies. This work is also required for insurance purposes. The outputs are often subject to third party audit by accredited risk engineering firms and insurance underwriters. Any changes to the design during FEED will require a revisit of the HAZID study to ensure that the risk register is accurate. In addition to providing compliance assurance, a benefit of the HAZID is to make informed design decisions and avoid major 'surprises' during commissioning.

The second round of process safety analysis usually involves the Hazard and Operability Study (HAZOP) study. This is a systematic review of the process plant to help identify hazards inherent in the process. Considerations include plant operability and maintainability, injury to personnel, asset damage, environmental damage, harm to the community, loss of production and liability. The key objectives include:

- Review of the final design for inherent hazards;
- Quantification of the risk associated with each hazard;
- Based on the level of risk, initiation of a process to ideally eliminate the hazard or reduce risk to ALARP.

The HAZOP, in simple terms, assesses the impact of the final design on operational practices such as maintenance and emergency management. For instance, during FEED, the volumes of hazardous substances within tanks and the processing facility will be determined by the design. From an operational perspective, this will inform the fire and explosion containment requirements for the site against mandatory standards. This will generate the need for emergency management plans, response capability and external agency involvement.

Setting an acceptable standard becomes more problematic for existing operations intending to undertake site or facility modifications that don't come under the scrutiny of the regulator through some sort of approvals process. For instance, a factory built 50 years ago may have a single lane access road. As the business grows, this access constraint would become problematic. For instance, due to lack of alternatives, it may end up being used by heavy vehicles entering and leaving the site as well as workers' and visitors' vehicles. This main access may also need to be crossed by workers when going to and from the factory to the main parking area. As a result, this interface between heavy vehicles, cars and pedestrians may become the highest risk on site.

Current legislation with its combination of 'WHS duty', 'reasonably practicable' and 'ALARP' doesn't necessarily provide a standard that prescribes the separations of heavy vehicles, cars, and pedestrians. Guidance material exists which may be taken into legal consideration if there were a fatality due to these traffic management risks, but there remains considerable latitude for the organisation to implement low-cost controls. The cheapest and least effective of these would be administrative controls (e.g. pedestrian crossing with give-way rules) and high visibility clothing (PPE) for workers traversing the area. In a culturally immature organisation, safety alone can be insufficient motivation for management to outlay the costs of new access roads using safety in design including separation principles.

Due to lack of clear regulation, oversight or intervention, ALARP can become symbolic rather than practised.

The limitations of ALARP may include a key misunderstanding that it's a formula with a go/no go answer at the end of the risk analysis in the design phase of new projects or modifications to existing facilities. ALARP tends to focus mostly on assets such as infrastructure, plant, and equipment, following the principles of ISO 55000 Asset Management, without the detailed inclusion of human factors. This can lead to the presumption that at the design phase the ALARP work is done. In mature organisations such as those following high reliability principles, the residual risk is revisited when there are major modifications to physical assets during the operational phase. For less mature organisations and more broadly in industry, even though ALARP has been set in concrete as an industry standard and a legislative benchmark; that doesn't mean that it should be accepted carte blanche. Practical testing should be used to validate the effectiveness of controls within the specific operational environment of the enterprise.

ALARP should be viewed with cynicism by operations who must live with design deficiencies and the risk of increasing rules based on low value thoroughness. The resulting compromised efficiency may generate new risks such as production pressure.

The danger with low value thoroughness is that it's time consuming, which in turn, reduces efficiency. Workers are then expected to get the work done in less time thereby creating organisational production pressure as well as individuals inflicting pressure on themselves to deliver on personal and team expectations.

Figure 3.4 has been developed to illustrate the ALARP Assumption using a new project as it transitions through various phases to operations. (There could be numerous versions of the concepts presented.) During the FEED Design Review Stage shown on the left-hand side, reliable risk reduction occurs because the HAZID and HAZOP focus on the higher order controls of elimination, substitution, and engineering. This is shown by the downward, dashed arrow.

The next phases together involve construction and commissioning. These are managed using safety critical controls for major hazards. In most instances, procedural controls and PPE for these phases are considered compliant and often to industry standards. In the figure, these are reliable resulting in a further reduction in the residual risk. As stated earlier, procedural controls and PPE such as permits to work and isolation procedures that are risk-based justify lowered residual risk. The arrow therefore continues downwardly.

During the initial phase of operations, what should occur is a focus on risk-based controls. The bulk of the OSH management system from a procedural perspective is developed and implemented for this phase. Provided these are risk-based, a further reduction occurs during the 'Operations: Major Risk Phase'. The figure shows that after this, the organisation moves to the 'Operations; Incremental Learning Phase'. Lessons come from incidents, internal audits and other opportunities for review. The rate of total risk reduction is shown to have slowed with the downward arrow being flatter.

As the operation and its OSH management system matures, there is the risk it will move into 'Operations: Low Value Thoroughness Phase', as shown. Actions are taken because 'something' must be done, and this may add greater attention to risk reduction with diminishing returns. Other risks with unforeseen or unacknowledged consequences may arise such as production pressure. The excessive use of rules may

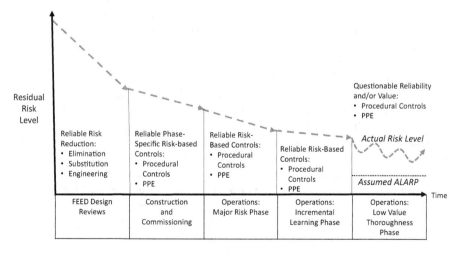

FIGURE 3.4   Project phases and the ALARP assumption

have negative impacts on the culture, for instance, if workers perceive that they're in a 'parent-child' relationship with 'the system'. For this reason, the actual risk level is shown as variable and unreliable. Below this squiggle is the line representing the ALARP level that management assumes has been achieved.

As an example, at some Australian mine sites, the controls for parking vehicles outside the administration area are rigorous compared to parking at a shopping centre. The standard controls include: the carpark is built on flat ground (safety in design the same as a shopping centre) and there are parking protocols that include leaving the vehicle in gear, with the park brake on and the keys removed (isolation of energy). Mine site vehicles are inspected by the driver regularly (daily or weekly) plus maintained according to the manufacturer's manual as a minimum. Controls that are additional to the standard controls often include: mandatory reverse parking if the vehicle is against a structure; and/or installation of humps over which the vehicle is parked so that if it rolls it will stop against the hump; and/or use of chocks. The additional controls are thoroughness without value given the rigour of vehicle maintenance and inspection regimes along with driver competency assessments. Being a highly visible area, compliance is easy to police. 'Policing' is the operative word in the example and will be discussed further in later chapters. In particular, the conversation will delve into the difference between visible enforcement and operational discipline as an outcome of risk management capacity.

ALARP, whilst accepted as the industry standard for risk reduction, has some serious deficiencies when it comes to operational implementation.

SUMMARY

Processes provide the interface between other systems of production. These are often proceduralised into four categories: those to describe how the system works; safe work practices; risk management practices; and incident investigation processes. There are various drivers that influence the development of such procedures and these include legislative requirements, standards such as ISO and self-regulation. From a practical perspective, processes bring technologies, the physical workplace and human resources together to achieve work. There are numerous activities across all industries where this happens daily. The compatibility of these systems at the interface has a significant influence on the risk level.

As a follow-up to Chapter 1 which discussed the Entropy Model, it was explained that risk is dynamic and worker exposure is a sum of a complex set of variables associated with systems of production. There are eight sources of risk comprising the total risk profile, which may contribute to losses including safety incidents. Human error fits within this profile. This was contextualised using the case studies of Deepwater Horizon and a train derailment. The latter, in particularly, highlighted that blaming the worker isn't a technically correct approach to incident investigation. In later chapters, 'blame' will be balanced against an appropriate level of accountability.

The concept of a Risk and Opportunity Principal Strategy was introduced and proposed that it may enable enterprises to adopt the principles of high-reliability organisations. This topic will be expanded progressively through the following chapters and will be the primary focus of the final chapter (Chapter 10). It was shown that senior managers should understand where their enterprise sits on the compliance continuum and if they're below the compliance level, the first objective should be to achieve this where risk tolerance is moderate (solely driven by compliance), the strategy is functional, and timeframes are based on a medium horizon. Proactive organisations on the other hand, have aspirational strategy, aim for sustainable profit, and work towards functional excellence. Entities that are reactive or pathological aren't only non-compliant but carry significant business risks that reflect the vulnerability of the enterprise and its officers.

This chapter has included discussion about some important definitions within the OSH legislative framework. The opportunity presented by the self-regulation pathway includes learning to perform well across functional areas. The concept of a Risk and Opportunity Principal Strategy was introduced and will be progressively expanded throughout this edition. There's nothing in current legislation that prevents safety compliance from also being efficient and contributing to quality work. The problem comes about from looking at this legislation only from an OSH discipline perspective.

Importantly, a sense-making framework approach is needed to develop safe, efficient, quality-driven work procedures. For these to be effective, they need to be orderly – simple where possible and if complicated, well-explained and socialised. Operations must be protected from chaos and complexity both of which introduce disorder. The role of support services professions, such as OSH, is to provide a buffer between sources of complexity and to provide the frontline with practical workable solutions. Sometimes those complexities are created by head offices that don't necessarily fully appreciate the challenges of operational implementation.

It was acknowledged in this chapter that the concept of ALARP and its equivalents are unlikely to change as these are the bases of risk management practices and current legislation. It should be viewed, however, with cynicism. Operations may inherit deficiencies in design because of risk-based decisions to contain project or modification costs. In addition, as a facility and its OSH management system matures, there is a danger of adopting further layers of lower order controls with diminishing return on risk reduction. Complexity may be introduced which causes risk to rise as workers become overwhelmed by bureaucracy and policing. It was proposed that risk transfer may be occurring as increased thoroughness, which is time consuming, results in unforeseen consequences such as production pressure and a negative impact on the workplace culture.

The next chapter will discuss technological residual and entropic risks, including transformations that are occurring within workplaces because of innovation. The following chapters (5 and 6) will discuss physical environment and human resources risks, respectively.

REFERENCES

American Institute of Chemical Engineers, Center for Chemical Process Safety, 2024: https://www.aiche.org/ccps/resources/glossary
Australasian Mine Safety Journal, 2013:
Safety driven. *Australasian Mine Safety Journal* (amsj.com.au).

Australian Broadcasting Commission, 2023:

BHP driver sacked after Pilbara train derailment claims unfair dismissal. *ABC News*.

BHP reaches settlement with driver sacked over iron ore train derailment in Pilbara. *ABC News*.

Handbrakes applied to wrong BHP iron ore train before Pilbara runaway derailment, ATSB finds. *ABC News*.

ATSB train derailment report raises concerns about BHP risk assessments and communication. *ABC News*.

Australian Mining, 2013, BHP accused of rolling over on vehicle safety. Australian Mining.

British Petroleum, 2023, deepwater-horizon-accident-investigation-report-executive-summary.pdf (bp.com).

British Safety Council, 2022, How Robens super-charged the safety system. British Safety Council (britsafe.org).

Carrillo, R.A., 2011, Complexity and safety – ScienceDirect. *Journal of Safety Research*, 42(4), 293–300.

CCPS (Center for Chemical Process Safety), 2009, *Guidelines for Developing Quantitative Safety Risk Criteria*. New York: Wiley.

Centers for Disease Control and Prevention, 2023, Radiation Studies – CDC. ALARA.

Foster, P., & Hoult, S., 2003, The safety journey: using a safety maturity model for safety planning and assurance in the UK coal mining industry. *Minerals*, 3(1), 59–72. https://doi.org/10.3390/min3010059 Article

Health and Safety Executive , 2023, Risk management: expert guidance. ALARP at a glance (hse.gov.uk).

Hollnagel, E., 2009, *The ETTO Principle: Efficiency-Thoroughness Trade-Off*. Boca Raton, FL: Routledge.

Hudson, P., 2003. Applying the lessons of high risk industries to health care. *Quality & Safety in Health Care*, 12, i7–i12.

Hudson, P., 2007, Implementing a safety culture in a major multi-national. *Safety Science*, 45, 697–722.

Levenson, N.G., 2012, *Complexity and Safety*. The MIT Press Cambridge, MA: SpringerLink.

Mary Kay O'Connor Process Safety Center, 2020, ALARP-Final-Paper-Publishing.pdf (tamu.edu).

NRMA, 2023, Changes to ANCAP safety criteria in 2023. The NRMA (mynrma.com.au).

National Institute for Occupational Safety and Health (NIOSH), 2023, Hierarchy of Controls. NIOSH | CDC.

Olewoyin & Hill, 2018, Chapter 29: Risk leadership – a multidisciplinary approach. In *Safety Leadership and Professional Development*. (Van der Stap), American Society of Safety Professionals (ASSP).

Sandone, G., & Wong, G.S., 2010, Making-Sense-of-Safety.pdf (cognitive-edge.com), Proceedings of the Association of Canadian Ergonomists 41st Annual Conference, Kelowna, BC, October 2010.

Stogran, Colonel (ret'd) Pat, 2016, Esprit de Corps Canadian Military Magazine, Understanding the military planning process — espritdecorps.

Van der Stap, T., 2008, Overcome the Conflict Between Safety and Production with Risk Management and Behavioural Safety Principles, *RM/Insight*, 8(2), American Society of Safety Professionals, USA.

WorkSafe Western Australia, 2021, The health and safety duty of an officer - interpretive guideline (www.wa.gov.au).

4 Technology

RESIDUAL RISK AND DESIGN

Technological risk management begins in the design, planning, and procurement stages during the initial phases of a facility's lifecycle. The external forces shown by the Channel affect these decision-making processes and encourage the enterprise to act responsibly, for example, by providing and maintaining, as far as reasonably practicable, working conditions in which workers are not exposed to hazards. As discussed previously, this is moderated against cost considerations. What is meant by 'does not pose a hazard' is that the level of risk is 'acceptable', given the current state of technological knowledge in the industry. It's not possible to eliminate residual risk, so all technologies are hazardous to some degree depending on the amount of energy involved.

OSH legislation imposes duties on manufacturers and suppliers to factor safety into designs, and to consider the risks to installers, those who undertake maintenance and operators/users of equipment. Requirements include testing and inspection regimes, as well as provision of information about the equipment, its hazards and safety during the product life. Duties apply to employers in relation to the use, cleaning, maintenance, transportation and disposal of plant, and the use, handling, processing, storage, transportation and disposal of substances. In addition to general duties, prescriptive regulations apply to plant and substances. These requirements tend to be clear cut.

'So far as is reasonably practicable' in the legislation alludes to two constraints that affect the level of residual risk of new technology and the rate at which it degrades. Firstly, at the design and manufacturing stage, risk-based cost-benefit decisions must be made concerning the level of safety built into a unit of equipment. For example, how much would a consumer be prepared to pay for a car that was 'guaranteed' to prevent serious injury in a collision at 100 kilometres per hour under set conditions? How much time and cost would a company be willing to dedicate to research and development of such a car? Limited resources cause the community to accept a level of risk that can't be reduced in the short-term. Organisations must manage these residual risks, which means in turn that workers will be exposed to some degree, for instance, during operations and maintenance.

Design-based, cost-benefit decisions also affect the life of a technology. Inputs that lengthen life of product are more expensive. The willingness of the consumer to pay for this extended utility influences product quality and the expected rate of degradation. Both residual and entropic risk levels of plant, equipment and tools are determined during manufacturing and relate to quality standards. As organisations are provided with known inputs when they purchase these traditional technologies, the entropic risk should be forecastable and measurable if the technology is operated

DOI: 10.1201/9781032701561-5

according to design tolerances. In this regard, these risks should be in the realm of known knowns – information is readily available to the purchaser and others in the industry.

The second constraint is that operations must make decisions about resourcing in the presence of scarcity. Human, physical and financial capitals are limited with little flexibility to increase these resources in the short-term. For example, the decision to replace old technology (physical capital) requires financial capital. With limited finances, the organisation may need, in the short-term, to accept the higher levels of residual risk inherent in the old compared to the new equipment. It may also have to compensate for higher rates of entropic risk using more rigorous, planned maintenance strategies. These constraints are well-understood in industry, but, as discussed, later in this chapter, there are trends towards significant changes in technologies that won't be so predictable and may take the enterprise into more unknown risks.

Properly designed and well-manufactured technology has a lead-time before it begins to degrade and ultimately reaches an 'unacceptable' risk. Ahead of that state, from the cost point-of-view, it may be beneficial to continue to operate the old technology in the short-term even though it's less efficient, less safe and more costly in terms of maintenance than the new technology. Such decisions can be made on two bases. The first is that the technological risk is still considered tolerable and that strategies are in place to manage this risk. The second is that the benefits of use continue to outweigh the costs. This should factor in the probability of the risk causing an incident and/or the losses resulting from sub-optimal productivity. There may or may not be hidden costs depending on what the organisation is measuring. The soundness of such decisions depends on the risk-based assumptions that have been made. A precautionary approach should be taken as organisations may gloss over the risk of failure.

The problem may be that factors that predispose the organisation to failure may not be measured simply because entropic risk prevention isn't part of the language of current organisations.

If it can be accepted that technologies suffer from degradation and if this is left uncorrected, a failure will happen, then there should be a more concerted effort to monitor degradation. Some large organisations are doing this using big data analytics, but smaller businesses either don't have access to the required resources or awareness of the need and consequences of inaction. Corporations are reporting the benefits with the biggest opportunities coming from the ability to predict what will happen, instead of recording what has happened or is happening (BP, 2014).

Technologies with significant levels of entropic and/or residual risk potentially have several negative impacts on productivity and safety. These include equipment downtime. Assuming the technology remains operational, degradation can lead to sub-optimal output, for example, an electric shovel (large-scale mining equipment) won't operate as efficiently if its hydraulic components are badly worn. In turn, this places higher demands on the operator to achieve the same output despite mechanical constraints.

Degradation of technology may also lead to sub-optimal interaction with the physical environment. As a simple example, if the blade of a grader is damaged, then the roads can't be maintained to the required standard, which, in turn, could become problematic for vehicles driven on these roads. At the strategic level, the costs of degradation are not limited to those that are easily quantified such as loss of machine availability. There are also the non-accounted costs of sub-optimal output and increased demands on other systems of production.

MANAGING TECHNOLOGICAL RISKS

The Entropy Model can be used as a reference point for decisions related to risk, mechanical tolerances and operability. It's common practice to push additional output or machine availability in production environments, but when does this become risky? The predictable behaviours of technologies are probably the easiest system of production to gauge changes in entropic risk. Concerning triggers include breakdowns, lack of redundancy, reduced production output and other production key performance indicators (KPIs).

The critical issue strategically is when should management push the alarm button if there are concerning trends in operational losses such as breakdowns and lower than expected production output, bearing in mind that the next loss could include a fatality or serious injury.

The nature of risk is that in most circumstances, the organisation will continue to perform but not necessarily to the standard expected. In addition, just because a situation shifts towards unsafe margins, it's not certain that an injury will be one of the losses incurred. The losses may be limited to production downtime due to technology issues. An event may pass without even a near-miss being recorded as such classified events require a judgement regarding proximity of a person. In practice, it's easy enough to state that the person was faraway enough to be out of the 'line of fire' and therefore not at risk. This would save time and effort required to undertake a safety investigation.

From a risk management perspective, an organisation that relies on its technology for production output should be preoccupied with failure. This fixation is one of the principles of those enterprises considered to be high-reliability organisations (HROs). HROs achieve nearly error-free operations over long time periods and maintain high levels of safety whilst operating in complex, uncertain, high-risk and hazardous environments (Roberts, 1990). An HRO has the capacity to anticipate and prevent undesirable and unexpected events by focusing on the proactive mitigation of risk rather than on responding to errors after these occur. If an unanticipated event occurs, it has the capability to firstly, cope with and contain errors from intensifying and secondly, restoring system functioning (Weick et al., 1999; Weick and Sutcliffe, 2007). Effective HROs apply five principles. These are:

1. Preoccupation with failure
2. Reluctance to simplify

3. Sensitivity to operations
4. Commitment to resilience
5. Deference to expertise

In this chapter, the focus will be primarily on the preoccupation with failure as it relates to technological risks; however, the other principles will be explained progressively throughout the rest of this edition, and detailed analysis will be provided in Chapter 10.

Figure 4.1 has been developed using the Entropy Model to illustrate how this fixation may affect operational risk-based decision-making in relation to taking corrective action when technological degradation is identified. This figure is a version of the Entropy Model so follows the same steps as described earlier. The difference is that the primary focus is on technology in isolation from the other three systems of production. For this reason, the first column only includes technology. The level of risk is zero which is an unrealistic paradigm. Reality is shown in the next section where there is a level of residual risk. This is consistent in the short-term due to knowledge and financial constraints, for instance, the high cost of plant replacements and modifications.

As technological degradation occurs (the downward dashed line), the entropic risk rises (the upward solid line). In the original Entropy Model, this would continue until losses become inevitable. After incurring such loss/es, the organisation takes corrective action followed by maintenance practices to achieve optimisation. In the figure, the enterprise that has a preoccupation with failure is continually monitoring technological degradation, and takes corrective action ahead of the escalation of entropic risk. This is shown for the arrow pointing to the left indicating correction is brought forward to achieve control.

Maintenance is also more proactively driven by understanding the high risks of failure, as shown by the upward dashed line in the 'maintenance to defined

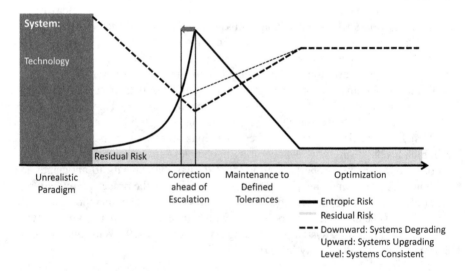

FIGURE 4.1   The fourth section of the Entropy Model with a shift in corrective action decision point

tolerances' phase. This would likely be more rigorous than simply following the original equipment manufacturer's instructions. The example illustrates an organisation that is concerned about the reliability and safety of its technologies.

By adding a principle of HRO behaviour – preoccupation with failure – the Entropy Model integrates production, safety, quality and reliability in a way that is readily understood operationally.

The same rationale may apply strategically. If senior management receives concerning data about operational performance across multiple functions, this should raise wariness to chronic levels. In HROs, this would certainly be the case. It's in relation to new and emerging technologies that the need for monitoring of degradation becomes even more important than for traditional, predictable, energy-based plant and equipment. These new technologies may introduce risks that aren't fully appreciated by society (or the enterprise) and therefore, the singular concept of 'risk', being the 'effect of uncertainty on objectives' needs to be refined.

Uncertainty has replaced probability and can be classified into two types – aleatory uncertainty and epistemic uncertainty. The former refers to inherent randomness or variability in a system or process, where the outcome cannot be predicted with assurance but can be described using probability distributions (Aven, 2016), like the roll of dice. Epistemic uncertainty arises from a lack of background information about an event, system, or process (Yoe, 2019; Aven, 2016). For example, if a new disease emerges and there is limited information about its transmission and mortality rates...

Taleb (2007) uses the term 'black swan' to refer to these unpredictable and unforeseen events that are rare and have a significant impact on society, organizations, or individuals. Aven and Krohn (2014) include these black swans as unknown unknowns in their categorization

(Aven, 2017; Lindhout et al., 2020)

1. *Known Knowns: The events that are already known and understood in risk analysis and confidently used in decision-making based on experiences or well-established facts.*
2. *Known-Unknowns: The events are known and are on the list of risk sources but are considered to have a negligible probability of occurrence and are thus ignored. Tsunami events were known in Japan, however due to judgement of an extremely low probability of occurrence were not considered a risk to Fukushima nuclear power plant (Aven, 2017).*
3. *Unknown Knowns: The events are not known to the analysts who conducted the risk assessment. They are either outside their realm of knowledge or they did not conduct a thorough risk analysis, but the events are known to others.*
4. *Unknown Unknowns: The events lie outside the realm of current knowledge, are unknown to the analysts and unknown to others. For example, a new and previously unknown virus outbreak, a sudden market crash, or a groundbreaking scientific discovery.*

(Van der Stap and Qureshi, 2024)

These four categorisations of risk are worth noting especially in the context of the Fourth Industrial Revolution discussed later in this chapter. Managing current technological residual and entropic risks shouldn't be problematic as these should fall into the category of known knowns for most organisations.

The discussions so far have implied that new technology is intrinsically safer than old technology. This may not always be the case plus it's possible for the reverse to be true in the short-term. The introduction of new equipment changes the risk to which workers are exposed because of interaction with other systems of production. Firstly, there is a negative impact on human resources. Current worker competencies may be inadequate to ensure safe, efficient operation using the new equipment. Personnel need time to become familiar with the change and this may cause degradation of the human resources system of production in the immediate term. Change control strategies must consider these risks by providing pre-operational training and practice to accelerate the learning curve, as well as additional support from supervisors.

Part of change management will likely involve the impact on processes and the need for new operational work procedures to be developed and implemented, in consultation with those who'll do the work. The physical workplace may need modification, for instance, additional ventilation, climate control, vibration protection or security measures. The procurement function should be consulted to manage new supplier relationships (such as service agents) and spare parts inventories. A risk assessment process must be part of the change management strategy; however, in many businesses, this is either expedited or overlooked completely.

CASE STUDIES: TECHNOLOGICAL DISASTERS

Relationships with major suppliers of technology can be a significant part of the risk management strategy for organisations that are heavily reliant on plant, equipment, and infrastructure to deliver core business. This should go beyond the simple provision of information as described in OSH legislation.

There should be greater transparency between customers and suppliers to prevent technological failures and to generate learning industries.

There are numerous examples where the lessons came because of regulatory investigation after the fact. It would be expected today that the management of these known knowns should be a shared responsibility amongst the various stakeholders in industries. The following three aviation incidents are an example of root cause commonalities that if better communications capability had been available at the time should have prevented the second and third incidents.

The first crash occurred in 1999 in Australia. The pilot of a Beechcraft Super King Air suffered hypoxia. The Australia Transport Safety Bureau (ATSB) recommended an aural warning to be fitted to these aircraft. They also proposed that authorities issue a directive for an immediate check of the fitting of passenger oxygen system mask container doors on all such aircraft and, all other aircraft with similar equipment (Wikipedia, 2024, 2000 Australia Beechcraft King Air crash – Wikipedia).

The second occurred in 1999 when a Learjet 35 business jet was flying from Florida to Texas. The aircraft lost cabin pressure and the six people on board were incapacitated by hypoxia. The incident is mostly remembered because the golfer, Payne Stewart, was onboard. Fighter jets were sent after the plane and the pilots noticed no structural damage. The windows, however, were frosted indicating the temperature inside was well below zero degrees (Wikipedia, 2024, 1999 South Dakota Learjet crash – Wikipedia).

Later, on 4 September 2000, a similar type of aircraft was involved in another disaster (Wikipedia, 2024, 2000 Australia Beechcraft King Air crash – Wikipedia) and was referred to as the 'ghost flight'. The findings of the investigation were inconclusive, but evidence suggested hypoxia. The likely causes were depressurisation of the aircraft and lack of aural warning system. For this third incident, the ATSB recommended aural alarm systems and low-cost flight data recorders for all pressurised aircraft. The earlier incidents should have prompted greater attention to these matters at the time. These incidents identified design deficiencies which more broadly, points to the need to include supplier relationships in the enterprise's risk profile. With the availability of worldwide communications, these relationships and information sharing should be readily achievable.

Other historic incidents highlight the need for risk assessment beyond the gates of the operation. For instance, there should be regular review of the dynamics outside a facility. The risk profile of an entire site can change with time in relation to its surroundings. An example is the explosion of a fireworks site in Enschede, Netherlands, on 13 May 2000 (Wikipedia, 2024, Enschede fireworks disaster – Wikipedia). Twenty-three people were killed and a further 950 injured when a warehouse storing fireworks exploded near a residential area. The original permit to store these goods was granted in 1977, when the surrounding area was not so heavily populated, and subsequently renewed three times, without a review of the changing risk profile surrounding the facility.

The most notable example of these infrastructure-related technological risks contributing to a major disaster is Bhopal. On the night of 2 December 1984, in Bhopal, India, a gas leak from a small pesticide plant owned by Union Carbide Corporation led to the immediate death of 2,259 people and injuries to a further 574,366 (Wikipedia, 2024, Bhopal disaster – Wikipedia). The incident resulted from the largely unrealised dangers associated with the manufacture of the highly toxic chemical, methyl isocyanate (MIC). The immediate cause of the MIC leak was an influx of water into the chemical storage tank but there were several catalysts adding to this including drought, agricultural economics, and the mismanagement of entropic and residual risk factors. One of the most significant factors was an excessively large inventory of hazardous substances causing the inherent dangers to be extremely high. The severity of the mismatch between the technological design of the plant and storage conditions with environmental factors therefore largely determined the extent of this disastrous event.

It's become evident from these discussions that the traditional approach to building a risk register based on physical hazards is grossly outdated. 'Hazards' are a narrow risk type within the broader context of the impact of the organisation and its operations on the community. This becomes an even greater field when considering product safety, the operational environmental footprint and the further reaching considerations within the ESG Agenda.

The granular risk detail remains important, but the bandwidth of risk-related professions has broadened and requires a 360-degree perspective vertically and horizontally, especially in response to changes in technology.

THE HUMAN–TECHNOLOGY INTERFACE

There's extensive literature on the human–technology interface including the field of ergonomics. The critical issues include fit between operator and machine along with exposure time associated with ergonomic risks. Some of the key areas have included use of computers, impacts of sedentary work and software driven monitoring programmes. Some of these are for the proactive benefit of the user such as scheduled breaks and exercises, whilst others may be thinly disguised productivity monitoring.

The 'quality' of either system has an impact on the other. For instance, poorly maintained plant puts additional demands on the operator, whilst an aggressive operator will increase the wear and tear on plant brakes and gears. The condition of the physical environment adds further risk variables to the person–machine interface. Process design is the final system of production to consider. For instance, a large mine truck will be designed to carry a certain tonnage capacity. The procedure may set some parameters for load limits and correct loading; however, there will be variability in the execution of this practice. The same applies to heavy vehicles on public roads where the issue is probably more problematic in under-regulated jurisdictions because of the desire to reduce costs of transporting goods and materials.

The challenge is that many of the risks of pushing the tolerances of technologies are unseen to the operator plus, in the context of production or schedule pressures, these may be marginalised in the decision-making process. A list of additional risks can be found in this South African Mine Safety article (Mining Safety, 2023) and includes:

- The vehicle will be less stable, difficult to steer and take longer to stop. Vehicles react differently when the maximum weights which they are designed to carry are exceeded.
- Overloaded vehicles can cause the tyres to overheat and wear rapidly which increases the chance of premature, dangerous and expensive failure or blow-outs.
- The driver's control and operating space in the overloaded vehicle is diminished, escalating the chances for an accident.
- The overloaded vehicle cannot accelerate as normal – making it difficult to overtake.
- At night, the headlights of an overloaded vehicle will tilt up, blinding oncoming drivers to possible debris or obstructions on the roadway.
- Brakes have to work harder due to 'the riding of brakes' and because the vehicle is heavier due to overloading. Brakes overheat and lose their effectiveness to stop the car.

- With overloading, seat belts are often not used as the aim is to pack in as many persons as possible into the vehicle.
- The whole suspension system comes under stress and, over time, the weakest point can give way.
- By overloading your vehicle you will incur higher maintenance costs to the vehicle – tyres, brakes, shock absorbers and higher fuel consumption.
- Insurance cover on overloaded vehicles may be void as overloading is illegal.

(Mining Safety, 2023)

Studies have shown the effects of road and truck conditions on driver vigilance using Chaos Theory (Hocking and Thompson, 1992). Different scenarios can be compared. For instance, firstly, if the vehicle is in a poor condition and the road is in a fair condition, results showed stable oscillations in the driver's vigilance. The operator varied in attentiveness without any serious consequences and was able to manage the demands associated with the poor state of the vehicle. Secondly, the vehicle was in good condition and the road in bad repair. The effect on the operator was chaotic oscillations in vigilance and the possibility of an accident. This situation put very high demands on the driver and the inconsistency of these demands made it more difficult to make the necessary adjustments, for example when the road has a pothole or uneven surface. The risk of an accident was contained by the driver's alertness, but this depended on their physical condition and competencies. The third scenario involved both the vehicle and road in bad condition. Chaotic oscillations in vigilance occurred and fell below the cut-off point where an accident was certain. The study lends weight to the Entropy Model and rising entropic risk, if left uncorrected, resulting in loss, which could include a safety incident and vehicle damage.

A table was provided in PSM1 of the strategies to manage risks associated with technology. It's repeated here as Table 4.1 to clarify the interfaces with other systems of production. These are where the complexities and dynamic nature of risk become evident, as described in the previous driving example. Since the publication of PSM1, there have been significant improvements in engineering which would support the examples of operational risk mitigations provided. There may be others that haven't been included.

The table separates strategies for residual and entropic risk management. Under residual risks, there are four main areas. The first is design and manufacturing short-comings of a technology. As discussed earlier, cost becomes a factor which determines the residual risk of a piece of plant or equipment. At the procurement phase, the enterprise should evaluate health, safety, environment, and quality (HSEQ) features of technological options as well at reliability and performance histories. Relationships should be established with suppliers to ensure that up-to-date information is available on the selected product. Once operational, the technology's performance should be monitored using data analytics.

The remaining three sources of risk are at the interface between the technology and each of the other systems of production, as shown. The fit between vehicle and road for instance is important, and this involves design and planning of the physical environment. When installing new equipment, risk factors and potential hazards should be evaluated and undertaken in accordance with manufacturer's instructions.

TABLE 4.1
Technological Risk Management Strategies

Source of Risk	Risk Management Strategy
	Residual risks
Design and manufacturing short-comings	• Pre-purchase evaluation of HSEQ, reliability and performance of technology options • Using supplier relationships to obtain up-to-date information • Monitoring of technology's performance using data analytics
Technology/Physical environment interface	• Effective design and planning of work environment • Pre-installation evaluation of risk factors and potential hazards • Installation of technology in accordance with manufacturer's instructions • Development of workplace-specific safety standards to manage residual risks
Technology/process interface	• Evaluation of suitability of technology for the process • Standardisation of technology • Modifications of process in the short-term to manage residual risk • Modification of technology to reduce residual risk in the longer-term
Technology/human resources interface	• Purchase of ergonomically-designed, operator-friendly technology • Modification of existing technologies to fit the operator • Job design • Modification of work practices with adaptability to human needs • Major hazard critical controls for the prevention of fatalities, injuries and debilitating illnesses
	Entropic risks
Wear and tear	• Proactive scheduled maintenance • Regular monitoring of condition of technology • Planned replacement of parts/equipment • Reactive maintenance
Technology/operator interface	• Pre-operation training • On-going refresher courses where necessary • Organisational culture which reinforces desired operator behaviours • Systems to correct undesirable behaviours

Thereafter, there should be specific workplace safety standards for managing these residual risks.

Technologies must be selected that are suitable for the processes or tasks required to be undertaken. Where possible, multiples of the same or similar plant and equipment should be purchased to exploit economies of standardisation. In the short-term, some processes may need to be modified to document the safe, efficient use of new equipment. Over the longer-term, plant modifications may reduce the residual risk.

Finally, the management of residual risk at the technology – human resources, again starts with procurement to optimise the fit at the interface. Modifications may be required to suit the individual operator, or if not possible, the job may need to be adapted to reduce exposure time, for instance, to whole body vibration when operating a piece of mobile plant. Where the technology has residual risks with fatality, serious injury or debilitating illness potential, these risks must be managed through major hazards or fatal risks critical control programmes.

The table shows that the two areas of technological entropic risk are firstly, wear and tear, and secondly, the technology-operator interface. The former is countered through proactive scheduled maintenance, regular monitoring of the equipment, planned replacements of parts and reactive maintenance. The fit between the operator and plant to prevent degradation is best managed through pre-operational training, refresher courses where needed, a culture of looking after equipment and correction of inappropriate operation such as supervision or coaching.

In today's business environment, the risks from supply chain and procurement are critical inputs to operational risk management and the organisation's capacity to sustain safe, efficient production.

According to McKinsey and Company, there are two concerns within the supply chain – absolute shortage risk and supplier risk. They explain that the risk of the former means products are unavailable in the quantities needed, irrespective of the price offered (McKinsey & Company, 2022). The latter involves individual suppliers defaulting. The procurement risk involves inflation where the products remain available but become more expensive. This is less of a concern for professions such as OSH but needs to be understood from the perspective of overall competition for scarce resources and the potential impact on the enterprise's total risk profile. For instance, higher costs of spare parts may be compensated for by a cut in the training budget.

There are strong commonalities between the procurement and OSH functions, regarding their relationship with production. Both serve to support core business by contributing to continuity of operations. In an article about procurement by McKinsey and Company, they advise on the need for early warning systems ahead of the next disruption. The procurement strategy should be on a continuum with 'continuity risk' being the highest level of concern and 'no immediate danger' being at the other end.

STANDARDISATION

Standardisation is a widely practised approach to managing risks and has provided numerous opportunities at the person–machine interface. For example, there are various studies across industries including the medical profession.

> Standardization of operative equipment can result in a significant cost reduction without impacting quality or delivery of care. Based on average case number per year, a

total annual cost savings of > $41,000 could be realized. Survey participants agree that standardization improves cost and patient safety…

(Avansino et al., 2013)

The focus of standardisation in production environments involves the procurement of same types of plant and vehicles to have a fleet for which the maintenance procedures, holding of spare parts and the required skills of mechanics become consistent. This, in turn, increases efficiency and safety based on internal capability. Oil and Gas customers are using the supply chain to streamline procurement using consistent equipment specifications to reduce facility costs and schedule delays (Postema and Wish, 2019). The Joint Industry Programme 33 has been established to make this step change improvement for the industry by having a standard approach to procuring equipment including for safety (IOGP, 2023). The automotive and aviation sectors have had such processes in place for some time and it will be interesting to see whether this efficiency strategy has wider uptake in the future.

Regardless of organisational size, improvements to safety, quality and efficiency can be achieved through simplification and standardisation. For instance, on a farm, operations can benefit from as many types of plant and equipment as possible being diesel-operated rather than having to store both diesel and petrol. There are efficiency (and safety) benefits from this approach.

More opportunities for improvement can be achieved by taking an operational perspective to workplace risk assessments than solely looking at safety, especially where capital investment is required.

Not surprisingly, general managers are more receptive to investing in recommendations when reported in a way that directly relates to core business. For instance, at a poultry farm, the manager and supervisors were asked what their greatest frustrations were. One of the biggest issues was that because rats often ate through the electrical wiring, the lighting in the poultry sheds regularly failed. The cabling had been installed continuously which meant that if one small section was damaged, the whole system for the shed went out. This was a significant issue for production, which was how it was reported to the general manager by the safety manager who did the site visit. It was agreed that the system would be rewired in several series to ensure if one section failed, the rest of the system would be intact. All wiring would be installed correctly in solid conduit. This resolved a very serious safety matter. Workers regularly washed down the sheds and it was possible that water could contact exposed electrical wires, causing electrocution. Tactics should be applied to achieve safety results through practical, operational improvements that directly impact core business.

MAINTENANCE OF TECHNOLOGY

The Entropy Model highlights the important of maintenance of systems of production. The links between the reliability of plant and equipment with safety and efficiency are irrefutable. The challenges from business to business in sustaining technological quality can be significant. Companies face different constraints and therefore their objectives will also vary. For instance, in some multinational mining companies that are well resourced and have large-scale operations, the purpose of maintenance is to ensure high levels of output and safety in the medium-term. These businesses may change technologies, including fleets, regularly to pursue economies of scale. Concurrently, they're potentially also reducing residual risks due to improvements in safety in design with next-generation plant, and exposure to entropic risks when the support services (including supplier relationships) are available to maintain the equipment.

In contrast, smaller-scale operators can struggle. Scarce financial resources often force these businesses to get much longer life from technologies thus increasing their vulnerability to entropic risk. The purpose of maintenance in their situation is to extend the operational productivity of existing equipment. Such companies may need to be more proactive to manage higher residual risks and to counter the degradation of aging plant and equipment. If not careful, this can create a vicious cycle of production time, followed by downtime for maintenance, before returning to operational productivity.

In all organisations regardless of scale, the over-riding issue is control. This involves maximising planned maintenance proportional to reactive maintenance. In the Entropy Model, reactive maintenance of technology is effectively a corrective action, but where there are trends of recurrent issues, this informs proactive scheduling ahead of critical safety margins and inefficiencies.

Rasmekomen and Parlikad (2014) discuss a generic approach to optimising maintenance of multi-component systems with degradation interactions. It's likely that large organisations already have these complexities under control from an operational and safety risk management perspective. In their paper, the researchers reviewed the work of Dekker et al. (1997) which breaks dependencies between components in multi-component systems into three types:

- Economic dependencies covering group maintenance and opportunistic maintenance to influence maintenance costs;
- Structural dependencies where components in the system are functionally bonded so that maintenance requires dismantling or concurrent upkeep of various connected parts;
- Stochastic dependence where failure or degradation of a component can influence failure of or degradation of others (Dekker et al., 1997).

Importantly, optimisation of maintenance can be calculated for components which degrade and can influence failure of other components based on the data related to the deterioration. In the paper, the researchers looked at a petrochemical plant and the interdependencies of gas tubes feeding excess-heated gas into a cold box unit. They were able to optimise maintenance thresholds. The inference for operations from studies such as these is that data analytics are allowing organisations to have greater understanding of what constitutes safe and reliable, and the point at which technology or its interconnected technologies, become unsafe and unreliable.

Performance-based maintenance of machinery is common practice in advanced industries such as oil and gas. The process relies on detection and prediction of performance degradation (Zagorowska et al., 2020). Interestingly, existing mathematical modelling of this performance assumes decreases at a consistent rate. They indicate that some performance degradation can reverse itself such as when a turbomachine self-cleans deposits on the blades under some conditions. The researchers suggest a dynamic 'moving window approach' to monitor technological degradation rather than assuming consistency. Referring to the Entropy Model in Figure 4.1, this suggests there could be fluctuations up and down along the degradation downward line and therefore, the entropic risk may also fluctuate. The overall principle of rising entropic risk however, remains sound.

A further area of technological risk management is whether to keep maintenance practices in-house or to outsource this work. Use of contracted services can go through various cycles of decentralisation (to outside) to centralisation (in-house). Historically, this has been a contentious issue particularly where the workforce is unionised as they prefer in-house servicing to retain members' jobs. The aviation industry has had challenges with the establishment of foreign repair stations. This has presented dilemmas for the Federal Aviation Administration in the USA and called for collaboration between international regulators. Olaganathan et al. (2020) undertook a study titled *"Managing Safety Risks in Airline Maintenance Outsourcing"*. They found that risks were complex and included:

- Lower standards of competencies, inefficiencies, deficiencies in the work environment and limited access to the internet and communication facilities;
- Cyber risks that pose a threat to safety such as electronic documentation and personal records;
- Differences in regulatory risks; and
- Use of unapproved parts.

The study found no strong evidence that using foreign repair stations to maintain the US carrier aircraft affected safety (Tang and Elias, 2012). It was believed that based on the study, if stakeholders work collaboratively and use maintenance data effectively, it will increase efficiency and reduce costs (Olaganathan et al., 2020). Another study done earlier came to a different conclusion:

> The concluding section highlights the potential latent failures arising from poor maintenance practice associated with outsourcing of aircraft maintenance to uncertified and poorly regulated shops.
>
> *(Quinlan et al., 2013)*

There are a few lessons from the example above even though foreign outsourcing by the US Air Force is well beyond the realms of smaller-scale organisations in the USA and elsewhere. Some of the notable risks are likely to apply even to SME's looking to outsource the maintenance of their mobile production plant (e.g. trucks, water carts, graders etc). For instance, if this work is done off site, then the owner of the plant has little or no control over the work environment, the competencies of the tradespersons undertaking the work or the maintenance service providers in-house record-keeping.

*Outsourcing infers a transfer of risk but from an OSH legislative
perspective, the expectation is that due diligence has been undertaken
prior to the engagement of contractors including service providers.*

The other risk factor worth noting is that the plant owner won't necessarily know if the maintenance service provider has used genuine manufacturer's parts or after-market parts. This may or may not affect the residual risk of the plant nor affect future wear and tear; however, this should be verified as part of the owner's or operator's OSH duty. For larger organisations that intend to have an ongoing relationship with the maintenance service provider, a review of some type should be conducted. This may range from asking for tradespersons' competency records and licences to operate, through to a site visit and audit. The action taken depends on the risk and local legislative requirements regarding 'due diligence' when ensuring WHS duties are met.

The technological risks described above are generally known knowns. The new and emerging technologies associated with the Fourth Industrial Revolution aren't so predictable. These are discussed in the next section.

TECHNOLOGY IN THE FOURTH INDUSTRIAL REVOLUTION

According to the World Economic Forum, the Fourth Industrial Revolution is a 'fundamental change in the way we live, work and relate to one another' (World Economic Forum, 2023). They state:

> There are three reasons why today's transformations represent not merely a prolongation of the Third Industrial Revolution but rather the arrival of a Fourth and distinct one: velocity, scope, and systems impact. The speed of current breakthroughs has no historical precedent. When compared with previous industrial revolutions, the Fourth is evolving at an exponential rather than a linear pace. Moreover, it is disrupting almost every industry in every country. And the breadth and depth of these changes herald the transformation of entire systems of production, management, and governance.

> *(World Economic Forum, 2016)*

Britannica provides a useful summary of key areas of change:

> It's important to appreciate that the Fourth Industrial Revolution involves a systemic change across many sectors and aspects of human life: the crosscutting impacts of emerging technologies are even more important than the exciting capabilities they represent. Our ability to edit the building blocks of life has recently been massively expanded by low-cost gene sequencing and techniques such as CRISPR; artificial intelligence is augmenting processes and skill in every industry; neurotechnology is making unprecedented strides in how we can use and influence the brain as the last frontier of human biology; automation is disrupting century-old transport and manufacturing paradigms; and technologies such as blockchain, used in executing cryptocurrency transactions, and smart materials are redefining and blurring the boundary between the digital and physical worlds.

> *(Schwab, 2023)*

* CRISPR is clustered regularly interspaced short palindromic repeats within gene sequencing (CRISPR Service - Creative Biolabs (creative-biolabs.com))

Interestingly, a second use of 'entropy' is being applied within this revolutionary process to the area of information theoretic. The original concept was based around Shannon's entropy and information which was developed from 1948 to 1951. This is the foundation of modern cryptography – the process of hiding or coding information so that only the audience it is intended for can read it (Kaspersky, 2023). Entropy is used to measure the probability of events and is particularly looking for anomalies to assess the collective change in probabilities. For instance, when mathematicians are studying language for artificial intelligence, they may detect neurological changes in people's conversation but instead of using traditional techniques, they don't know what they're looking for.

The older style of thinking about the world is to look for knowns, measure it and/or fix it. In the new style, the assumption is not knowing what to look for.

What is now more important is to look for unusual trends.

This concept is important in complex systems such as those that use artificial intelligence (AI) because of the need to measure unseen and unknowable risks. In the past way of measuring old known known technological risks, a measure was developed to predict failure and if the variable moved beyond a certain value, an intervention is taken. With AI, scientists are looking for tiny changes that indicate something may be going wrong without having to wait for large amounts of data to support this conclusion. What is inferred is that businesses should be looking for matters that are out of place when these new technologies have been implemented.

There have been several applications of the mathematical approach to Entropy Theory to evaluate OSH. For instance, Dunwen et al. (2012) applied it to building construction safety. They stated:

> To improve the safety management level of construction site, a new kind of safety evaluation method using the combination of fuzzy mathematics and the entropy theory was put forward...The results show that the hybrid method, which can overcome the shortcomings of each single method, is feasible, practical and operational in construction site safety assessment.

(Dunwen et al., 2012)

With the guidance of Dr Andrew Back, advance research fellow in artificial intelligence, from the School of Electrical Engineering and Computer Science, University of Queensland, Australia, the following five key industrial trends were identified. These are visualised from an operational risk and safety perspective based on the author's interpretation of what the future might look like.

TREND 1 – HUMAN–MACHINE COOPERATION

Future work will involve people working cooperatively with machines. This is increasingly common today in support of people with impairments and in this regard, there is extensive literature. What is unknown is what this will entail in workplaces of the near future and what the safety and productivity risks will be. People tire but machines don't so how will the pace of work be determined? Perhaps the human's eyes will be monitored for fatigue but the other subtleties such as stress, may be the real issues. This must be considered given that OSH legislation has recently included psychosocial harm as a workplace risk that has to be managed by the organisation.

It's also accepted in the OSH profession that people will work around rules and deviances can become normalised. If a machine works in a predictable way such as automated passenger vans that take personnel to different areas of the site along known routes at specific times, then there is a risk that human drivers will make judgements about 'safe' distances for passing in front of automated vehicles. Rules may be implemented but, for instance, a 20 second rule, may be interpreted as a slow 20 seconds or a quick 20 seconds (if left to people to count) unless that's also measured.

The question is how the risk assessment will be done ahead of the introduction of technological changes, and how will human behaviour be predicted including deviances. As a simple example, drivers already make such judgements when approaching traffic lights that turn amber. Do they stop or proceed? Many factors go into this instantaneous decision including speed, distance, other nearby vehicles, the traffic ahead (whether it's moving or not) and if a police vehicle is in the vicinity. If a person is working with a machine that operates in a predictable way, there will be behavioural choices available that may push safety margins. The use of monitoring may be needed to ensure compliance, but this would be a significant regression in OSH management practice which is centred on a culture of trust.

Trend 2 – Balance of Power in Human–Machine Systems

The second related trend comes about because humans and machines communicate differently. Humans relate on both task and social bases which isn't set to a consistent pattern but happens as needed or spontaneously. Working with a machine would be an impediment to this so the issue is, will the human have the power to 'pause' the machine or will humans have to adapt/ suppress their natural tendencies within this hybrid work environment. Perhaps, the precursory conditioning will have occurred through the current generations' close relationship with their mobile phone.

In a broader context, there are likely to be issues related to locus of control. People have no control over the technology or the disruption from this major transformation (Xu et al., 2018). For instance, as machines are complex systems designed by engineers which are then imposed on the target work group, there may be poor fit because of different perspectives of the work between the engineers who design the machines and those who work with the machines. This adds a whole new dimension to 'Work as Imagined' and 'Work as Done'. All the procedural controls will follow the engineered model, perhaps eliminating the need for consultation with worker. Consultation and participation are currently corner stones of 'enabling' OSH legislation based on duty of care principles. Workers may feel that they have less and less control within the workplace. Historically, changes occurred gradually, and workers had time to adapt, whereas the near future demands a different competency set focused on flexibility and transferability of capabilities including to unknown contexts and learning environments.

Some jobs will be threatened by automation, but human work overall will remain central to organisational performance. Routine jobs with a high volume of tasks related to information exchange, sales, data management, manual work, construction and office work are more exposed to the risk of automation (Grau and Wang, 2020). In this regard, as the transformation progresses, it will be interesting to review data on the impact across countries and by industry within countries. The World Economic Forum provides regular articles on latest developments (World Economic Forum, 2023).

TREND 3 – NONLINEARITY OF COMPLEX MACHINE ADOPTION

Traditionally, risk has been reasonably predictable because systems of production have behaved in linear ways. For instance, the rates of degradation of technology and infrastructure can be predicted to some extent, given a few known variables such as throughputs and hours of operation. Even human degradation has been studied such as physical and mental fatigue over time. Ill-health by type has a degree of predictable impact on employee absenteeism. This knowledgebase has built up with volumes of data analysed statistically.

The fourth industrial generation is much more complex. In the discussion earlier, the concept of stochastic dependence was raised. This was where failure or degradation of a component can influence failure of or degradation of others (Dekker et al., 1997). There may be unknown risks within these new highly complex systems such that failure of one component could break the entire system. If this is the case, then either the system will be fully usable or relegated to the scrap heap.

Highly complex systems could have unknown unknown risks that lead to total failure. The problem is that traditional approaches to risk assessment won't detect these risks.

This presents significant challenges for operations utilising systems of production. If an operation becomes highly automated and the system goes down, for instance a fleet of autonomous trucks at an open pit mine site, will such trucks be capable of being operated manually? How will the company plan for a significant failure? It would be impractical and cost prohibitive to have a standby fleet with human drivers at the ready. This scenario highlights that 'maintenance' will have new dimensions – traditional physical plant and equipment plus process instrumentation maintenance as well as AI systems. It seems that these complex systems and the interconnection of components and things, will introduce a different type of economy of scale that will reap tremendous benefits if the system works as predicted, but potential for major failures if it doesn't.

Earlier discussions talked about new projects and the residual risks and potential entropic risks inherited by operations because of risk-based, cost-benefit decisions made at the project planning stage. The Fourth Industrial Revolution is likely to bring technologies into the workplace that aren't readily understood by operations.

What these technologies produce will be visible in some way whether by inspection or from data analytics, but the 'mechanism' of the system will be the realm of elite specialists. In that regard, it's easier to accept that most people who drive a car don't know about the transfers of energy that make it work, nor will the end-user or co-worker to the machine need to know how it operates, only what it does or can do.

Trend 4 – Trust between Agents and Reliability of the Whole System

The fourth trend is a highly complicated issue of trust. How ready is society, from a values perspective, for these new technologies and AI, in particular? The fundamental question is to what extent will this new industrial revolution serve humanity – what value will it bring. The highly unpredictable nature of this transformation is likely to make sectors in society nervous.

It's uncertain at this time, how operational risk will be assessed when these new technologies are introduced to the workplace. Will workers 'comply' with the system or create workarounds that undermine it? The availability of social media means that a single voice of dissent can become a wider campaign of resistance. Traditional language associated with management of change will likely become redundant. How will new human-machine operations be 'socialised'? Will change be done in a test environment or done cart blanche, given issues such as the pressure on management to act quickly or miss a perceived opportunity? As discussed in Chapter 3, some major corporations tend to set industry standards which then becomes an expectation of other organisations including contractors. The challenge is that once the standard is set, it becomes 'general knowledge' and captured within the enabling clauses of OSH legislation's 'duty of care' or 'WHS duty'.

Many traditional technologies were able to be tried and tested on a small scale before implementation across operations. The potential investments involved in new technologies may be difficult to revert from once management's commitment has been made. Alternatively, it may be that fears of change and loss of control will be exaggerated and humans will continue to be self-serving but with a better appreciation of global impacts as the ESG Agenda grows concurrently. Some commentators have highlighted that there needs to be a balance of industrial progress with social progress within this fourth industrial revolution that was deficient in previous revolutions (Mulgan, 2021).

Trend 5 – Privacy vs Efficiency

A fifth issue will likely be cyber security trends. A current ethical conflict is the demand for privacy and security, which is now becoming at odds with the way in which systems are being built. An example is current software which monitors how people work in the office environment, which is likely to extend to factories and other workplaces. The push for greater efficiency may be met with resistance against being measured and observed. This is already problematic with the use of CCTV. An example was provided in earlier chapters of the traditional production line where co-workers failed to assist an injured worker due to production pressures and lack of worker control over the pace of the production process. In this case, CCTV was a clear visual means of monitoring, whereas new technologies will be increasingly covert.

> *The monitoring of people will likely create a greater divide between management and workers.*

> *The risk to safety and efficiency may be systems that work but are introduced for the wrong reasons or have counter-productive consequences.*

BROADER ISSUES IN RISK-BASED DECISION-MAKING

An article in *Risk Management Magazine* (2022) discussed unintended consequences of new working models that resulted from the COVID-19 pandemic, including the 'Great Resignation'. Businesses have struggled since to manage their talent risk and maintain an organisational culture. There appears to be a trend towards increasing enterprise risk appetite, but this presents a quandary as in one survey it was found that 69% of significant business decisions were not risk informed. Shrinkman states:

> Organizations face potentially steeper consequences from such decisions. According to Gartner's analysis, nearly one in three decisions at the senior executive level can have an adverse effect on key stakeholders, and 19% of decisions made at this level could invite increased regulatory scrutiny.

> *(Shrinkman, 2022)*

In the context of the Fourth Industrial Revolution, the pace of change means that Executive and Senior Managers won't be able to wait for more data to make decisions. It's suggested that a different approach is needed which involves 'sense-making'. This has four key elements:

1. Synthesizing risk information from disparate sources and providing a weight to the information to establish context;
2. Prioritising the most relevant risk information to make key considerations more obvious;
3. Helping decision-makers apply the information in the organisational context;
4. Preparing decision-makers to make sense of risk information independently.

> *(Shrinkman, 2022)*

Along with the growth in AI for beneficial purposes, are rapidly evolving malicious AI actors and cyber attackers (Wagner and Furst, 2018). This is a risk that resides at the enterprise risk management level, but the consequences for operations and support functions such as OSH and environmental management should also be considered. The questions include: If the system fails, where will people be, and will this put them in danger? What response plans need to be in place for different system failure scenarios? How will contingency plans affect the various functions of the organisation?

Finally and perhaps cynically, the OSH profession has spent the last 20 years (along with other functions), driving the need for worker input and feedback to reach management. This is a cyclical process where input from operations results either in actions taken by senior management or at least, a response regarding why action might not be feasible. This is a pragmatic approach to continual improvement. With the rapid advancement of AI, how will this top-down bottom-up communication continue?

How will workers raise legitimate concerns about the functionality of AI in the workplace? Will it be trivialised because it's assumed that AI is right?

Will management be receptive to 'dumb' questions (which may in fact be the best to ask)?

SUMMARY

Current and future technologies have been the focus of this chapter. Despite significant changes occurring in the industrial world, there remains the need for more effective management of residual and entropic risks to pursue continuity of production, quality outputs and HSE performance. The fundamentals of HSE in design, quality of componentry and standards of construction and commissioning are critical inputs to the facilities that operations inherit. It was explained that these project planning decisions determine the residual risks and potential entropic risks that need to be mitigated during the operational phase of a facility's lifecycle.

One of the significant issues that was raised is that organisations may have a predisposition to failure because entropic risk prevention is not part of the language of operations. The solution provided was the adoption of the five principles of HROs, in particular, preoccupation with failure. This should provide the impetus for greater transparency across organisations and industries using global communications capabilities, to learn and act in the interests of preventing disasters. This is urgent with the rapid emergence of Fourth Industrial Revolution technologies that will have unknown unknown risks.

Whilst the OSH and other risk-related professions have made solid progress in the last 20 years to create alignment with other disciplines such as engineering, HRM, quality and environmental management, that's insufficient for this new more complicated world. A 360-degree perspective scanning vertically and horizontally will enable such professionals to better appreciate supply chain complexities and the influence of other stakeholders (both internal and external) on the future of their work. This includes a better appreciation of enterprise level risk-based decision-making affecting technological choices. Some of the key themes were standardisation, the broadening of what constitutes technological maintenance and whether such work is done in-house or outsourced.

The Fourth Industrial Revolution is transforming the human-technology interface. This is likely to add greater dynamism and intricacies. Traditional man-machine fit was based on ergonomics but in future, it also must include psychosocial risk factors such as locus of control. Much of the progress in the OSH and HR professions in the last 20 years has been centred on increasing levels of consultation and participation in operational risk management and continual improvement more broadly. The impact of AI could pressure these strategies, putting more faith in AI reliability than human innovation and initiative.

The final area that was discussed in this chapter was the limitations of traditional enterprise risk management practices. As raised by other commentators, there's a definite need to shift towards 'sense-making' as the foundation of risk-based decision-making, especially in the context of increasing unknown unknown risks as mankind ventures into untested technological realms.

The following chapter will continue the exploration of the systems of production by looking at residual and entropic risks associated with the physical environment / workplace, before presenting the final chapter of this series on human resources risks.

REFERENCES

Avansino, J.R., Goldin, A.B., Risley, R., Waldhausen, J.H.T., & Sawin, R.S., 2013, *Standardization of Operative Equipment Reduces Cost*. ScienceDirect.

Aven, T., 2016, Risk assessment and risk management: review of recent advances on their foundation. *European Journal of Operational Research*, 253(1), 1–13. https://doi.org/10.1016/j.ejor.2015.12.023

Aven, T., 2017, A conceptual foundation for assessing and managing risk, surprises and black swans. In G. Motet, & C. Bieder (Eds.), *The Illusion of Risk Control* (pp. 23–39). Cham, Switzerland: Springer.

Aven, T., & Krohn, B. S., 2014, A new perspective on how to understand, assess and manage risk and the unforeseen. *Reliability Engineering & System Safety*, 121, 1–10. https://doi.org/10.1016/j.ress.2013.07.005

British Petroleum, 2014, Number crunching with Big Data. News and insights. Home (bp.com).

Creative Biolabs, 2023, CRISPR service - creative biolabs (creative-biolabs.com).

Dekker, R., van der Duyn Schouten, F.A., & Wildeman, R.E., 1997, A review of multi-component maintenance models with economic dependence. *Mathematical Methods of Operations Research*, 45, 411–435.

Dunwen, L., Lei, Y., & Bo, L., 2012, Fuzzy-entropy theory comprehensive evaluations method and its application in building construction safety. *International Symposium on Safety Science and Engineering in China ISSE*-2012, *SciVerse Science Direct. Procedia Engineering*, 443, 137–142.

Grau, A., & Wang, Z., 2020, *Industrial Robotics – New Paradigms, IntechOpen, Industrial Robotics: New Paradigms.* Google Books.

Hocking, B., & Thompson, C.J., 1992, Chaos theory of occupational accidents, *Journal of Occupational Health and Safety - Australia and New Zealand*, 8(2), 99–108.

International Association of Oil & Gas Producers, 2023, *Safety Archives*. IOGP Publications Library.

Kaspersky, 2023, What is cryptography? (kaspersky.com).

Lindhout, P., Kingston-Howlett, J., Hansen, F. T., & Reniers, G., 2020, Reducing unknown risk: The safety engineers' new horizon. *Journal of Loss Prevention in the Process Industries*, 68, 104330. https://doi.org/10.1016/j.jlp.2020.104330

McKinsey & Company, 2022, Managing supply chain risks. McKinsey.

Mining Safety South Africa, 2023, Vehicle overloading. Mining Safety.

Mulgan, G., 2021, The social economy and the fourth industrial revolution (ssir.org).

Olaganathan, R., Miller, M., & Mrusek, B.M., 2020, *Managing Safety Risks in Airline Maintenance Outsourcing*. Managing Safety Risks in Airline Maintenance Outsourcing (erau.edu).

Postema, A., & Wish, D., 2019, Standardization of procuremen, A.t equipment specifications: a step change in facility cost and schedule reduction. OTC Offshore Technology Conference. OnePetro.

Quinlan, M., Hampson, I., & Gregson, S., 2013, *Outsourcing and Offshoring Aircraft Maintenance in the US: Implications for Safety*. ScienceDirect.

Rasmekomen, N., & Parlikad, A.K., 2014, Optimising maintenance of multi-component systems with degradation interactions (sciencedirectassets.com). *IFAC Proceedings*, 47(3), 7098–7103.

Roberts, K.H., 1990, Some characteristics of one type of high reliability organization. Organization Science, 1(2), 160–176. doi:10.1287/orsc.1.2.160

Schwab, K., 2023, The Fourth Industrial Revolution. *Encyclopedia Britannica*, 31 May. 2023. https://www.britannica.com/topic/The-Fourth-Industrial-Revolution-2119734. Accessed 28 February 2024.

Shrinkman, M., 2022, Risk management magazine – helping leaders make risk-informed decisions (rmmagazine.com).

Taleb, N., 2007, *The Black Swan: The Impact of the Highly Improbable*. New York: Random House.

Tang, R., & Elias, B., 2012, Offshoring of airline maintenance: implications for domestic jobs and aviation safety. *Congressional Research Service*, 1–30. Retrieved from https://www.researchgate.net/publication/298106135_Offshoring_of_airline_maintenance_Implications_for_domestic_jobs_and_aviation_safety

Van der Stap, T., & Qureshi, Z., 2024, Can organizations achieve high reliability and manage risk more effectively using the entropy model?, Unpublished work at the time of writing.

Wagner, D., & Furst, K., 2018, Risk management magazine - artificial intelligence and risk management (rmmagazine.com).

Weick, K.E., & Sutcliffe, K.M., 2007, *Managing the Unexpected: Resilient Performance in an Age of Uncertainty*. San Francisco, CA: Jossey-Bass.

Weick, K.E., Sutcliffe, K., & Obstfeld, D., 1999, Organizing for high reliability: process of collective mindfulness. In B.M. Staw, & R. Sutton (Eds.), *Research in Organizational Behavior* (vol. 1, pp. 81–123). Greenwich, CT: JAI Press.

Wikipedia, 2024:

1999 South Dakota Learjet crash – Wikipedia.

2000 Australia Beechcraft King Air crash – Wikipedia.

Enschede fireworks disaster – Wikipedia.

Bhopal disaster – Wikipedia.

World Economic Forum, 2016, The fourth industrial revolution: what it means and how to respond. World Economic Forum (weforum.org).

World Economic Forum, 2023, Fourth industrial revolution. World Economic Forum (weforum.org).

Xu, M., David J.M., & Kim, S.H., 2018, The fourth industrial revolution: opportunities and challenges contents (sc-celje.si).

Yoe, C., 2019, *Principles of Risk Analysis: Decision Making Under Uncertainty*. Boca Raton: CRC Press.

Zagorowska, M., Spuntrup, F.S., Ditlefsen, A-M., Imsland, L., Lunde, E., & Thornhill, N.F., 2020, *Adaptive Detection and Prediction of Performance Degradation in Off-Shore Turbomachinery*. ScienceDirect.

5 Physical Environment

RESIDUAL RISK AND DESIGN

The physical environment (PE) includes infrastructure and locational factors such as the natural features of the site, including topography, flora and fauna, climate, and ground conditions. The risk assessment for this system of production should be done on four levels. The first is the immediate work environment where the task is being undertaken. The second is the site level whilst the third includes the nearby surroundings of each operation covering neighbouring properties and an understanding of the risks associated with the area. For instance, a vehicle service facility will have risks within the site but if an explosives factory is 500 metres away, that risk should also be a concern in case of an incident and the need for local evacuation. The final level is regional which takes into consideration the potential for natural disasters such as floods, cyclones/hurricanes and bushfire.

The organisation is legally required to manage the risks associated with the physical environment, as far as practicable, to prevent injuries to workers and other persons on site. This obligation extends to potential consequences like injuries to the public and damage to infrastructure and natural systems located outside operational boundaries.

The Channel described in Chapter 2 explained that the enterprise is required to be both legally compliant and socially responsible. In most developed countries, there are stringent controls to ensure that organisations operate conscientiously to prevent negative impacts to the environment within the operational site and beyond the gates. It's in this area of risk control that there is a strong relationship between OSH and environmental management.

Historically, the connection between the two functions has been well-accepted. Case studies undertaken in the Netherlands and Denmark indicated that OSH and environmental management are linked. On a technical basis, this stems from the notion that hazard sources are similar and therefore these programmes have synergies (Kamp and Le Bansch, 1998). When managing PE risks using an OHS management system, this may concurrently contribute to the prevention of environmental damage. Likewise environmental management strategies can have a positive impact on health and safety, for instance, emission controls provide safer atmospheres for surrounding communities as well as workers. In the past, there was some reluctance to integrate these disciplines. For instance:

> Much stronger than is the case for OHS management, in environmental management there is an emphasis on external control, a requirement of supply chain management, and of responsibility for product effects during their entire life cycle. As a consequence,

DOI: 10.1201/9781032701561-6

the material topics that are dealt with in environmental management may have stronger links to all pervasive risks – and thus to strategic company issues – than those on OHS management.

(Kamp and Le Bansch, 1998)

As discussed previously, the concept of integrated management systems has now widely been accepted, with this reinforced by the commonalities between the ISO Standards 45001 (safety), 14001 (environment) and 9001 (quality). From a cultural perspective, there are robust alignments with the focus on mitigating losses, and in the case of HROs, preoccupation with failure. The Entropy Model with its central theme of loss prevention includes inefficiencies, safety incidents, quality deficiencies and environmental harm. The opportunity for seamlessness of these functions is attributable to the common driver of risk management.

OSH, environmental and quality management are also linked through the relationship between organisations and their suppliers. In Chapter 4, it was suggested that stronger interdependencies should be developed involving information sharing to better manage risks by reducing product/ technological residual risks and the tendency of such goods to degrade. Through mutuality in the supply chain, there are more options to improve product development by responding to customer feedback and achieving higher standards of technology and workplace designs. Such improvements should contribute concurrently towards minimising negative environmental impacts of the organisation's operations. OSH and environmental management both focus on maintaining the quality of the PE system of production to varying degrees.

These disciplines whilst overlapping must maintain respective specialist knowledge to prevent one management system being relegated to the other. Conceptually, a set of boundaries is needed to minimise role conflict and to promote a multi-disciplinary approach with high degrees of expertise together with collaborative problem-solving. In Chapter 4, a second principle of HROs that was briefly discussed was 'deference to expertise' in challenging times. A Venn diagram approach may be useful to clarify responsibilities for worker safety versus environmental impacts. For instance, control of groundwater runoff and leaching to the water table are unlikely to have a direct impact on worker health and safety. If however, the runoff creates a site hazard, then there will be the need for risk mitigations in both areas. Any uncertainty regarding roles and responsibilities can be addressed using a collaboratively developed HSE / EHS / Sustainability Management Plan.

The Enschede and Bhopal disasters discussed in Chapter 4 were examples where the surrounding communities were impacted by the respective company's operations, not just the workers. There has been a long history of disasters where having operations in challenging environments has led to major incidents. These include mining operations such as the Esmeralda cyanide spill in Romania in 2000 (ABC News, 2000), the Brumadinho upstream tailings dam collapse in 2019 (ABC News, 2019) and others as described by Piciullo et al. (2022). The researchers found that a lack of understanding of foundation ground conditions was a factor in several tailings dam failures. There were foundation problems resulting from lack of proper site investigation prior to construction. The unknown risk caused by a weak foundation subsoil went undetected. This was exacerbated by deficiencies in the upstream and downstream construction methods.

In the paper, there were numerous variables related to tailings dam construction type. It should be borne in mind though that when it comes to the PE, more generally, there may be the risk of 'black swans' (unknown unknowns) and incomplete data. For instance, geophysics analysis of ground conditions is based on sampling methods. There is also a problem of ambiguity in geophysical interpretation and many different geological configurations could reproduce the observed measurements (Keary et al., 2002). These types of uncertainties lead to further research to better understand unforeseeable risks. For instance, Duong (2023) discusses analysis of earthquakes and the risks these seismic events have on international tailings dam failures. There are monitoring regimes in place with stability assessments every two years and advancements now include improved strength estimations in the event of an earthquake. This involves assessing dam wall saturation levels against the risk of collapse.

When it comes to natural events such as earthquakes and cyclones/hurricanes, the risks are uncontrollable for organisations as there will always be limited information available to make risk-based decisions. This was the case for instance, with the tsunamis that hit Japan in 2011 even though the power plant operators knew there was a risk but failed to appreciate the severity or likelihood (Stanford News, 2021).

Physical environment risks can be difficult to assess because most 'what-if' scenarios will lead to the conclusion that the event could happen, but whether it warrants the resources for prevention becomes a contentious issue for decision-makers.

These types of occurrences tend to result in reactive responses when needed at the time of potential crisis. It's therefore important to consider where the organisation stands legally and ethically when disaster strikes to better understand the continuum of accountability – uncertainty – inculpability. The discussion below raises some considerations.

'ACTS OF GOD'

The first issue is that risks that are beyond the control of the organisation, such as one in 100 year flood occurring more often than predicted, fall into the category of residual risks. Such risks according to the Entropy Model must be managed in the short term due to technological and financial constraints. In the longer-term, the risk may be reduced through investment in better safety and environment in design. Such investment needs to occur according to the area of impact. Natural disasters need to be planned at the regional level whilst at the granular workplace level, it's the responsibility of the site's management. Overall, however, the principle of fit applies.

Risk management focuses on creating a better fit between man-made systems and natural systems to prevent failures.

The second consideration is when does an event become an 'Act of God'. This is where a standard of reasonableness must be applied. Diverting for a moment: there is currently debate in the OSH profession about the avoidance of 'blame culture' (Sherratt et al., 2023). Commentators are suggesting that a 'no-blame culture' is leading to lack of accountability and a growth in incidents being categorised as 'Acts of God'. This becomes problematic because almost every area in systems of production can be traced back to human input, whether that's design, manufacture, maintenance or operation. Ascribing 'no blame', in the interests of fairness, can't then only apply to workers. It must apply to engineers who design workplaces and technologies (such as tailings dams) and the managers who determine resources for maintenance, monitoring, and other risk management practices.

The problem is that without accountability, there's no learning from failures or successes. Calling an incident an 'Act of God' needs to be justifiable.

The question that emerges is: Should society accept that there is such a thing as an 'Act of God' and if so, using what criteria? The answer must be risk-based. Undoubtedly, this is a topic for an entirely separate book which some researchers have aimed to address already, for instance, Faure et al., 2014–2015; Mercante, 2005–2006; and Swanson, 2013.

The test of 'Act of God' is most likely a legal one done on a case-by-case basis and therefore, the best advice is not to rely on this as a defence should the organisation suffer a major disaster. Where it becomes more complicated according to Schwartz (2020) is the legal doctrine of 'Force Majeure' which specifies what should happen in case of an 'Act of God' like the coronavirus. In the paper, the author predicts changes to these contractual clauses to ensure coverage of pandemics but warns there may be some undesirable consequences from a business perspective including modified pricing strategies. Hansen (2020) did a study of the COVID-19 outbreak in terms of force majeure fundamental aspects and identified the following characteristics:

Unforeseeable – not reasonably foreseeable by parties when they are entering into the contract

Unavoidable – neither party could prevent the occurrence of the event or circumstance

Uncontrollable – incapability of contracting parties to control the event and its impact

Impracticable – the event and its impact have adversely affected the fulfillment of contractual obligations

Beyond responsibility – the event is not substantially attributable to a party

(Hansen, 2020)

These characteristics could well apply to all acts of nature or acts beyond human control. When used as test criteria, they help to better define the PE system of production residual and entropic risks which are within the control of the organisation. A simple, reverse logic can be applied.

If it's foreseeable, avoidable, controllable, practical and the organisation's responsibility, it's not an 'Act of God'. Every other undesirable event resulting from the natural environment, should be subjected to scrutiny on a case-by-case basis.

MANAGING PHYSICAL ENVIRONMENT RISKS

RESIDUAL RISKS

Initially, the management of residual risks should focus on those risks for which there's the potential for fatalities whether the source of risk is technological, the physical environment, human factors, the work method (process) or the foreseeable combinations of these. Companies describe these internal control requirements as 'major hazard standards', 'fatal risk protocols' or similar. These documented safety critical controls are usually determined by a risk assessment method such as bowtie analysis. Numerous internet sites provide an explanation of how this method works including available software to support the process, for instance, from the UK Civil Aviation Authority (2023) and NC State Poole College of Management (2016).

As a summary, the bowtie has the major undesirable event at the middle of the bowtie shape, for instance, ground collapse or flooding of the site. On the left-hand side are the preventative measures, some of which will be safety critical whilst others are non-critical. An example of critical is using seismic monitoring and data analysis to send an alert well ahead of ground failure. On the right-hand side of the major event, are the measures to mitigate the severity of the consequences. This will include having evacuation plans and rescue capability at the ready. The process then involves determining how often the critical controls need to be verified for adequacy and effectiveness. This is done through specific programmes such as inspection regimes or data gathering – whatever information provides evidence for the assurance process.

The approach to the development of standards for PE residual risks varies from country to country. In the USA, the Occupational Safety and Health Administration (OSHA, 2024) has developed standards by hazard type such as biological agents, earthquakes and wildfire. In the UK, the Health and Safety Executive embed PE residual risks within guidance material for specific industries such as diving, fire and rescue, mining, and offshore oil and gas. For instance, for safe havens in underground mines, the documentation covers thermal environment, air supply and environmental monitoring. Both OSHA and the Health and Safety Executive provide references to the relevant legislation. In Australia, there are extensive resources available for hazard management including ground/ strata instability, inundation and inrush, air quality and dust, gas outbursts, and ionising radiation (Australasian Mine Safety Journal, 2019).

One of the critical areas of PE residual risk management, at all four levels of risk assessment, is emergency response. Even in low-risk workplaces such as offices, evacuation plans need to be in place and rehearsed on a regular basis. Planning for response in case of emergency should be an inherent part of the work. This becomes crucial the further the workplace is, in terms of time and/or distance, from support services, such as medical facilities or emergency agencies. Remoteness is a PE residual risk best managed by advanced planning, regular re-evaluation, reliable transportation and communications.

The planning should centre on worst-case scenarios. When it comes to remoteness, the time required to reach medical facilities or for emergency services to be mobilised is the most important factor. This will be affected by PE residual risks such as terrain, road conditions and weather. The planning isn't something to be done annually or even monthly when workers are moving to different locations, which is the case with industries such as exploration, forestry, fishing, surveying, and other remote-based work. A review should be done as part of planning for the day's work. With natural environment residual risks, conditions can change so communications capability is essential. An example of what can go wrong occurred in September 2011, when marathon runners were trapped in a gorge in a remote area of Western Australia where a bushfire had taken hold. It took five hours before rescuers reached the two women who had suffered burns to about 80% of their bodies (News.com.au, 2011; ABC News, 2014).

Whilst regional emergency response is primarily the responsibility of government and various agencies, organisations should stay informed of those PE residual risks which have the potential to impact their operations. Information on trends is readily available on the internet, for instance, 'The Changing Risk and Burden of Wildfire in the United States' (Burke et al., 2021). Emergency alerts and status updates from agencies are often accessible via an app in real time or with minor delays.

At the local level, the managers of each enterprise should know the likely time for medical assistance to arrive at their site and the scope of potential injuries that could occur, for instance, suspension trauma from a fall whilst wearing a restraint harness. The medical conditions of individuals on site should also be known and planned for. The legislation of the relevant jurisdiction will likely stipulate the minimum requirements; however, care for injured or unwell workers should be the overriding driver of investment in response capability.

As discussed further in Chapter 6 on HR risks, it may be the case that some individuals, due to personal factors, aren't suited to work in certain physical environments, such as extremes of temperature, confined spaces, working at heights etc. If an organisation chooses to employ the individual, then management must ensure that the risks are managed according to that person's condition. A precautionary approach should be taken, if possible, to only employ workers who are fit for the PE residual risks associated with the job. In general, a programme of acclimatisation can assist personnel to adapt to conditions. As a further example, some organisations expect their representatives to travel overseas for work. This should be preceded by risk assessment and a personalised management plan for the traveller, including evacuation in case of emergency or threat.

Most of the discussion in this chapter so far has focused on the natural environment. Infrastructure also falls into this area as it constitutes a physical workplace. The residual risks of the human-made workplace are like those of technology. These relate primarily to the placement of workers in relation to dangers. Traditional workplace conditions readily come to mind such as floors, lighting, ventilation, access and egress but of more concern are unknowns and invisible risks, such as fumes and biological hazards. The World Health Organization 2017 reported on indoor environments and respiratory illness, including allergy in 1997 but coined the term 'sick building syndrome' back in 1982 (Jansz, 2017). Symptoms and signs can be categorised into two dimensions. These are anatomical sites such as eyes, respiratory tract, skin and central nervous system, and possible mechanisms such as irritation and inflammation, allergic reactions, toxicity, infections and environmental psychological stress. Guidance on evaluating buildings is provided by the United

States Environmental Protection Authority (1991) and numerous empirical studies have been completed since.

Unfortunately, unknown or unforeseen risks can be associated with infrastructure, as was the case with the Grenfell Tower fire in the UK in 2017. The fire was started by an electrical fault in a refrigerator on the fourth floor. It spread up the building's exterior accelerated by combustible cladding and external insulation with an air gap between which enabled a stack effect. Authorities, building owners, management and residents were unaware of this design deficiency. Seventy-two people died, 70 were injured and 223 escaped.

There has been regulatory change in the UK since the tower fire (UK Government, 2020). The US Fire Administration acknowledges the inherent nature of tall buildings and the need for safety in design to include fire-rated compartmenta-tion, non-combustible construction, fire alarm systems, automatic fire sprinklers, smoke control systems and multiple fire and smoke protected exit stairs (US Fire Administration, 2022).

Society accepts the operation of high-risk industries, but homes must be built for the safety of inhabitants and the community. High-risk industries are often those that have inherent dangers associated with the location where the work is undertaken. The residual risk sources can have an accumulative effect on the overall risk. For instance, in underground coal mining, risks include the obvious physical aspects of darkness, uneven surfaces, access and egress issues, air quality and water inrush. Hidden dangers can include toxic gases and unventilated areas, within a working environment which is basically combustible. Offshore rigs are considered high risk because of the flammability of the material being extracted, which is exacerbated by remoteness if a fire and/or explosion occurs (as was the case with the Deepwater Horizon Disaster).

The residual risk of both technologies and the physical environment can often be explained by simply considering the energies involved and the impact of loss of control of such energies.

The same energy principle applies to non-work environments and infrastructure so proximity of the energy source to populations is a major consideration in the Western World, but pressures in developing countries make locational factors more challenging.

The strategies for managing the residual and entropic risks associated with the PE system of production are summarised in Table 5.1. As is the case in the manage-ment of technological risks, the control of these risks begins in the design stage. In Western countries, it's mandatory for site assessments to be undertaken and the risks to be clearly identified as part of project planning and application for regulatory approvals.

Table 5.1 highlights that some sites, especially those that were built well before strict government environmental requirements, may have design shortcomings, for instance, the footprint may be too small for the storage of equipment making site movements more difficult. Ideally, the organisation would begin with a 'blank canvas',

TABLE 5.1
Physical Environment Risk Management Strategies

Source of Risk	Risk Management Strategy
	Residual Risk
Site design shortcomings	• Pre-development evaluation of physical environment characteristics and risks • Monitoring of impact of the physical environment on operational performance • Regular reviews of PE residual risks • Monitoring changes to the PE inside and beyond the operation's footprint
Physical environment/Process interface	• Evaluation of suitability of the process for the physical environment • Modification of existing processes in the short-term • Modification of physical environment (where possible) in the longer-term
Physical environment/Technology interface	• Effective design and planning of work environment • Pre-installation evaluation of risk factors and potential hazards • Installation of infrastructure in accordance with engineering standards
Physical environment/Human resources interface	• Modification of the physical environment to fit the needs of workers as far as practicable • Flexibility in work practices to keep the demands on the worker to an acceptable level • Major hazard critical controls for the prevention of fatalities, injuries and debilitating illnesses
	Entropic Risk
Natural degradation of the physical environment	• Regular monitoring of the condition of the physical environment • Emergency response procedures and contingency plans • Maintenance of natural environment and infrastructure
Physical environment/Process interface	• Regular monitoring of condition of physical environment • Safe work practices/operating procedures • Proactive scheduled maintenance of the physical environment • Reactive maintenance
Physical environment/Technology interface	• Regular monitoring of condition of physical environment • Proactive scheduled maintenance of the physical environment • Reactive maintenance
Physical environment/Human resources interface	• Regular monitoring of condition of physical environment • Good housekeeping practices • Organisational culture which reinforces desired behaviors

but this isn't always the case. The key strategy when planning changes is what the site will look like when all modifications have been completed. This is to prevent ad hoc changes. As an example, management of an established food processing operation built a new storage facility in the footprint of the old facility, then realised this hadn't fixed the issue of a constrained area where the greatest traffic movements occurred in front of the building. If management took a longer-term perspective, they may have located the new facility further from the production centre. In industry more generally, the failure to aim for optimisation of design can result in lost opportunities that hamper the enterprise for a long time impacting both safety and efficiency. Ad hoc changes without strategic planning can be problematic for businesses.

Adhocracy can be a cultural habit in organisations that don't have in-house experience and engineering competencies. As a result, safety and efficiency in design can be overlooked or poorly done.

At the PE-process interface, how the work is done should regularly come under review especially when work occurs in physically demanding conditions, such as work outdoors during summer months. Some basic changes could include doing more manually intensive work during the cooler times of day or organising inward supplier deliveries for early morning and outward deliveries for late afternoon through coordination with stakeholders.

The opportunity is to challenge why work is done the way it's done and how it can be executed more easily.

For instance, in remote exploration, one of the biggest challenges is the logistics of getting all the equipment including temporary accommodation, out to the field hundreds of kilometres away. This can be a time-consuming process for instance, using ten-tonne vehicle-mounted cranes to load materials onto the backs of trucks at the base facility and unloaded on site in 40°C heat. If trucks are supplied by a contractor, then issues arise because trucks aren't always available, and this option can be expensive. In this scenario, a supervisor challenged how the work was done asking if there was a better way. Management listened. The solution selected by the operations team was to purchase a truck which had a flatbed tray that could be unloaded using the onboard truck-mounted crane. A layout was designed so that all the goods were secured onto the tray at ground level at the base facility, and after mobilisation to site, the tray was unloaded and left on the ground using the crane attachment. At the end of the drilling campaign, the tray was simply reloaded onto the truck and returned to the base facility with all the equipment onboard as well as waste from site. Using cost-benefit analysis, the truck paid for itself within 12 months. Field crew already held truck operating licences, so the organisation was able to exploit previously under-utilised capability. Overall, the process was made safer, more efficient, flexible, and cost effective. The residual risks of working in a remote, harsh environment were taken into account and work processes adapted using a technological solution and in-house competencies.

An ethos was developed of saving time and energy. This can lead to more efficient ways of working. Workers were encouraged to challenge the status quo.

Having to work in extreme heat should be questioned, for instance: can we do the work somewhere cooler; or can we reduce the amount of time working in the heat; or can we reduce the level of exertion required to do the tasks in the heat? If people feel too awkward to ask in fear of seeming foolish, lateral thinking can be used, as explained in: The Most Valuable Skill In Difficult Times Is Lateral Thinking – Here's How To Do It (forbes.com) (Forbes, 2020). The principles of this simple example can be applied in any organisation to explore opportunities for improvement that lead to more effective management of risk.

As explained in Chapter 4, many risks occur at the interfaces between systems of production. Poor management of change can create operational risk issues. Even corporations can make the mistake of pursuing economies of scale in production without considering the consequences. For instance a mining company purchased a fleet of higher capacity ore trucks to discover that the bucket couldn't be tilted up in a maintenance workshop which was designed for the smaller fleet. There are numerous news stories of oversized lorries getting stuck under bridges due to poor logistics and route planning. These blunders can be avoided with effective management of change and by including risk assessment in the planning phase.

Residual risks also occur at the interface between the PE and HR. The fit between workers and the PE isn't static especially when the work is undertaken in various locations throughout the day. This demands more cognitive adjustments by the worker than being in a non-dynamic environment like an office or workshop. Compare for instance, running on a treadmill to doing a cross-country run. A dynamic relationship between the person and their environment requires a constant process of looking ahead, adjusting foot placement and avoiding obstacles – in other words, risk management.

A 'fixed' physical workplace can be modified to some extent to fit the needs of workers. Job arrangements under demanding PE conditions should be flexible to enable continuity of safe, productive work. Major hazard critical controls should be implemented and tested to prevent fatalities, injuries, and debilitating illnesses. In changing work locations, PE residual risk critical controls could include for instance, heat stress, bushfire, lightning and medical evacuation from remote locations.

What is apparent with PE residual risks is that it's not always possible to change the work environment. Such is the nature of residual risks so questions must be raised about how to make conditions more workable at the physical environment–human resources interface.

ENTROPIC RISKS

The PE is also subject to entropic risk as it degrades. The two types of physical environments discussed above included the natural environment and human-made infrastructure. Degradation of the latter is the same as technology and so the preventative strategies are also similar. Within the built workplace, deteriorated conditions often arise from human activity. In the OSH field, this has resulted in housekeeping standards being introduced in many workplaces. In the quality management field, it's been framed within the 5S of Lean Production.

> The 5S quality tool is derived from five Japanese terms beginning with the letter "S" used to create a workplace suited for visual control and lean production. The pillars of 5S are simple to learn and important to implement:
>
> - Seiri: To separate needed tools, parts, and instructions from unneeded materials and to remove the unneeded ones.
> - Seiton: To neatly arrange and identify parts and tools for ease of use.
> - Seiso: To conduct a cleanup campaign.
> - Seiketsu: To conduct seiri, seiton, and seiso daily to maintain a workplace in perfect condition.
> - Shitsuke: To form the habit of always following the first four S's.
>
> *(American Society of Quality, 2024)*

> *The English translation is to use the terms sort, set in order, shine, standardise and sustain. From the reference, the benefits are "improved safety, higher equipment availability, lower defect rates, reduced costs, increased production agility and flexibility, improved employee morale, better asset utilization, and enhanced enterprise image to customers, suppliers, employees, and management". (American Society of Quality, 2024)*

At a fundamental level, this involves modifying the workplace so there is a place for everything in its place. This is an area that the OSH profession has focused on as a measure of safety performance by undertaking workplace inspections, but this presents a contradiction to the reality of work.

The problem with 'safety' is that maintaining housekeeping standards all the time is inefficient and not how work gets done.

Often OSH documentation will refer to 'at all times' which simply isn't practical. To the 'outsider' such as the OSH Advisor who walks into a workshop to do a spot check on housekeeping, a work area may look chaotic with tools on benches, powered equipment like compressors wheeled into the active work area, and waste containers filled to the brim. What will the OSH Advisor report on the checklist? What score will be given for the area? Ask the workers how they might rate themselves and the response might be, "it's organised chaos". In a perfect world, tools get put away when no longer needed but it's human nature to keep things available in case they're needed again during the day. It's likely that it's more efficient to do it this way. The easiest way to consider what is an acceptable level of 'visual chaos' is to consider whether it supports efficiency or has reached a point where it gets in the way of efficiency. The other consideration is whether it's unsafe. Spills on the floor are unsafe because of the risk of slipping but clutter isn't unsafe. So what standard should apply?

Every workplace has its 'saturation point' where 'visual chaos' (as perceived by the outsider) transitions from supporting efficiency to becoming inefficient. Once this is reached, safety and quality are also compromised.

For example, in the process of creating magnificent meals for customers in the expected timeframe, a chef will accumulate pots, pans and all sorts of cooking tools turning the kitchen into a visual disaster area. Once the meals are served however, attention turns to getting the restaurant ready for the next main meal session. This will be the pattern of work with risk unless the business owner can afford to employ someone as a full-time kitchenhand who does all the cleaning progressively.

The same rationale applies in other workplaces. If the chef becomes hindered in their ability to produce output and meet customer expectations because of this chaos, changes need to be made. In this regard, it's more helpful to consider risk in relation to active production timeframes. For a restaurant, this may be the lunch session then the dinner session. For a 24-hour operation, this may be the three 8-hour shifts assigned to the production line. Rather than setting housekeeping standards and applying them 'at all times', it's more realistic to accept that workplace risks are dynamic, so some degradation is normal due to activity. The defined production cycle (shift or workday) should be streamlined by housekeeping standards that ensure operational readiness at the start of each cycle. This means that the workplace is restored for optimal safety, efficiency and quality at the end of the shift, in readiness for the next shift or workday.

Operational efficiency can drive the work, subject to prompt corrective action for those conditions that create a safety risk.

The focus of housekeeping standards should be operational readiness for the next production cycle not 'at all times'.

One reason for raising this issue is that OSH management systems often have standards and procedures, with broad statements like 'at all times' and 'all workers' etc. There's a gap between 'Work as Imagined' and 'Work as Done' because of the sheer impracticality of achieving this perfect world. 5S being a Japanese system of work which has strong cultural relevance, may be applied more as guiding principles in countries that are not as perfection and strict discipline oriented as a broader societal cultural trait. The 'at all times' pressure to pursue continual perfection may have a negative consequence of creating a psychosociological hazard. This is an area worthy of further research.

Housekeeping standards should be discussed from a risk-based perspective to support efficient operations not just safety. Unfortunately, these standards are usually branded as safety documents. The best question for OSH professionals to ask the operations and maintenance teams, is, 'What does operational readiness look like for your work area? What do you need to be safe, efficient and to do a quality job?' These are the changes that should be pursued, because at the end of the day or shift, the workplace should be restored to a state with minimal degradation, which, in turn, enables efficient, safe start-up in the next production cycle. This is a key mitigation in the PE–HR interface as shown in Table 5.1.

The other matter to consider occurs when more work is expected to be done within a footprint. This will put pressure on the PE and other systems of production and make the prevention of degradation more difficult. As stated earlier, every workplace has its 'saturation point' where chaos transitions from supporting efficiency to becoming inefficient. This is particularly the case where the number of workers increases beyond the original design capacity, such as in old factories. Another cause can be the introduction of additional lines of production within the limited workspace. At some stage, what tends to occur is safety incidents rise along with quality defects as systems are pushed beyond tolerances. The long-term solution is to bring the work output back to within design capacity of the facility, for instance, by building new facilities for additional production streams at another part of the site or another location.

There are numerous topics that could have been selected for the discussion about PE residual and entropic risks. The above has focused on those with significant operational relevance across industry types. The natural PE requires particular attention because climate change is making the environment less predictable and there appears to be a rise in unknown unknowns. Owners and managers need to be more outward looking at risks beyond their operation's gates.

THE WORKPLACE IN THE FOURTH INDUSTRIAL REVOLUTION

In Chapter 4, the discussions about technology turned to the Fourth Industrial Revolution and its impact on the world of business. This has also brought about changes in what constitutes a workplace and how future work environments will be designed. Some of the key trends include: the gig economy, autonomous operations, adaptation for an aging workforce and ecosystems.

There are numerous resources describing the current gig economy and its rate of growth. This trend is characterised by fragmentation which results in flexibility, shifting of risks to workers and income instability (Weil, 2014). There is a disconnected, individualised experience of work because workers aren't physically together in a common location. Two categories have been identified – 'crowd work' and 'work on demand' as explained by Kaine and Josserand (2019). Crowd work is generally done at the worker's home and can involve 'microtasks' or performance of professional expertise such as delivering work online. In this case, the person's home constitutes a workplace and how that's set-up from a risk perspective depends on that person's preferences, not necessarily on a risk basis. The second category includes tasks such as point-to-point transport, food delivery and services by means of private vehicle such as pet care. In these cases, the workplace is variable and often includes public spaces. Kaine and Josserand (2019) provide a summary of literature on gig work which is a good platform for further research.

A key issue with the gig economy is that responsibility for risks falls to the gig worker. The three main vulnerabilities are occupational, precarity and platform-based work. Occupational vulnerabilities come about because gig workers may have the same exposures such as increased risk of traffic accidents during deliveries or dangers of entering an unfamiliar home to provide care-giving services, without any regulatory protections (Bajwa et al., 2018). Precarity relates to the temporary nature of the work and the psychological distress this can cause as well as having to provide their own resources to perform the work. Finally, platform-based work has many considerations including isolation, unbundling of work and surveillance of the worker's task performance (Bajwa et al., 2018).

The principles of risk management of traditional fixed workplaces don't readily apply. This home-based or mobile work situation is quite a different scenario than the trend resulting from COVID-19 which has now allowed employees and direct contractors to work to onsite, remote or hybrid employment/engagement models. In this case, the employer still has a clear OSH duty even though they don't directly control the workspace, which is often the worker's home office or vehicle.

Another significant trend that ties in with technological innovation is the removal of some workers from the operational area. This is achieved using autonomous machinery such as driverless haul trucks in open cut mines or ore trains travelling from mine to port. The control room may be many kilometres away in a metropolitan area whilst the operations are in remote locations. The traditional workplace (e.g. the open cut mine pit) still exists but some categories of workers (the truck operators) are no longer required. According to Rezaei and Caulfield (2021), current views are that autonomous vehicles (AVs) may reduce the number of accidents but not necessarily the severity of any accidents. The researchers' study found that most industry professionals preferred that AVs still have steering wheels so that AVs could be manually operated, as a contingency. This answers a question which was raised in earlier discussions about such technologies and infers that skilled operators will still be needed on site where autonomous plant is used. With time however, as confidence grows in the safety and reliability of the technology, this may change.

As explained in a video from Transurban Group (Connected and automated vehicle trials | Transurban Group, 2023), the 'workplace', which in this case is public roads, needs to be adapted, for instance with clear line markings for the vehicles to work autonomously. At the time of writing, testing involved assessing how an automated truck interfaces with road technology; in other words, how the road will 'talk' to the vehicle (Transurban Group, 2023). From a risk management perspective, the scope for the physical work environment to have the capacity to communicate with machines (and perhaps people) is likely to be another opportunity to reduce residual risks and prevent entropic risks in the future.

The profile of human resources risks is also undergoing transition. There are significant economic issues ahead due to the aging population. The World Health Organization warns that retirement-age and older adults currently outnumber children under the age of five (Forbes, 2022). Interestingly, the term 'longevity economy' has been used. According to the US Bureau of Labor Statistics by 2030, all baby boomers (post WWII to 1964) will be at least 65 years old. 9.5% of the civilian labour force is forecast to be older than 65 years (U.S. Bureau of Labor Statistics, 2021).

As people will be working longer, the design of workplaces will need to be adapted to ensure productivity, safety, and health. Organisationally, it would be prudent to recognise the degradation of the workforce due to aging as an emerging risk factor. Key areas of human deterioration will need to be considered. These include physical strength, flexibility, balance, sight, reaction time and speed, hearing, manual dexterity and tactile feedback, body fat, respiratory and cardiovascular systems, systemic blood pressure, fatigue and ability to work in extreme temperatures (Perry, 2010). There are also psychosocial considerations including shift preferences and risks of disengagement. Perry (2010) lists design considerations that include processes, technologies and the physical work environment. Research continues to be undertaken to create a better fit between older workers and the workplace.

More broadly, management practices are evolving because of these and other social, technological, and economic pressures. According to an MIT SMR-Deloitte survey on the future of the workforce, most managers now consider employees and external workers to be part of their workforce (MIT SMR-Deloitte, 2021). The term 'Workforce Ecosystem' has emerged and is explained briefly at: Practical guide to workforce ecosystems | Deloitte Insights (2021) and MITSloan Management Review, 2021. Again, the physical concept of the workplace is under challenge which has implications for professions such as OSH, particularly if or when, regulatory requirements are updated to encompass this new reality of work. In Australia, changes have already taken place. OSH legislation has been updated to replace the term 'employee' with 'worker' and 'employer' with 'PCBU':

Worker
   Any person who carries out work for a PCBU, including work as an employee, contractor, subcontractor, self-employed person, outworker, apprentice or trainee, work experience student, employee of a labour hire company placed with a 'host employer' and volunteers.
   PCBU
   Person conducting a business or undertaking. The model WHS Act places the primary duty of care on the PCBU. The term PCBU is an umbrella concept used to capture all types of working arrangements or structures. A PCBU can be a: company; unincorporated body or association; sole trader or self-employed person. Individuals who are in a partnership that is conducting a business will individually and collectively be a PCBU.

(Safe Work Australia, 2023)

The definition of 'workplace' is worth noting when considering the above terms:

Workplace
   Any place where work is carried out for a business or undertaking and includes any place where a worker goes, or is likely to be, while at work. This may include offices, factories, shops, construction sites, vehicles, ships, aircraft or other mobile structures on land or water.

(Safe Work Australia, 2023)

When reading the above definitions through a lens that includes the gig economy, the longevity economy, and the workforce ecosystem, it would be difficult to argue that these new categories of worker, work and workplace don't fall within the employer/PCBU's OSH duty. This will be, or perhaps already is, a challenge for organisations to manage from a traditional health and safety perspective, which highlights the necessity for a risk-based approach. Strategies based on silos will be unmanageable during this Fourth Industrial Revolution because the boundaries of responsibility between the 'employer' and the 'worker' tread on the sensitive areas of privacy and security. The employer should hesitate to overstep these boundaries. Actions taken in the interests of the workers' physical health and safety, such as monitoring keyboard actions or active computer time to prompt breaks and exercises, could cause psychological distress, if remote workers perceive these interventions as 'Big Brother' watching. Psychosocial hazards and risk management are now included in OSH legislation in some jurisdictions, so the management of OSH risks has become increasingly complex.

SUMMARY

In this chapter, the discussion turned to the residual and entropic risks associated with the PE system of production. It was clear that like technological risks, HSE in design is the major mitigation. For the physical workplace, environmental protection considerations are also critical to minimise residual risks inherited by operations and potential entropic risks that will develop through the facility or site lifecycle.

Human-made technologies and workplaces should be built to known risks according to engineering standards then managed through maintenance and monitoring regimes to keep those risks within tolerances. The natural physical environment, however, can lead to 'black swans' due to unknown unknowns and incomplete data causing constraints to the risk assessment process. These unknown unknowns, once human interventions have been excluded, may be 'Acts of God'. It's important to clearly contextualise this concept, given the current trend towards 'no blame' cultures which may be eroding objective incident investigation and accountabilities.

In high-risk industries, extensive work has been done to develop and implement fatal risk controls including monitoring regimes and emergency response capability, particularly in the presence of these workplace residual risks. Other organisations also have potential exposures to fatal risks so it's important that business owners and managers understand their work and environment including regional risk exposures beyond the operation's gates. Current climatic changes make it imperative that businesses prepare for potential emergency situations and not just rely on government services.

In this chapter, discussions have included site design. Many old facilities suffer from sub-optimal productivity, safety, environment and quality plus miss opportunities for improvement through modifications because of bad management habits akin to adhocracy. Residual risks are reduced through long-term planning and good site design that supports operational efficiencies and safety. Many organisations would benefit from adopting a longer-term, strategic risk management approach to their current and future operations.

As with other systems of production, the workplace is subject to degradation, and this is most raised as a safety issue when assessing housekeeping standards. The discussion included contextualising the Lean Production Principle of 5S which involves having a place for everything and everything in its place. The reason this needs to be critiqued is that spot safety inspections can be unhelpful as work is dynamic and some degree of 'organised chaos' is part of efficiency. Safety critical risks with fatal or serious injury potential, however, shouldn't be accepted. It was explained that the physical workplace has a 'saturation point' where this 'visual chaos' transitions from supporting efficiency to becoming inefficient.

Finally, some forecasts were made of the future of the workplace emerging from the Fourth Industrial Revolution. The key issues identified included the definition of 'workplace' and how it's no longer simply a common office or site where everyone arrives Monday through to Friday to work in a shared environment. Significant changes have occurred due to the gig economy. The need for change continues in response to an aging workforce, the impact of human-less interaction between the built environment and technologies, and the emergence of workforce ecosystems that demand a more mature risk management approach.

In the following chapter, the residual and entropic risks of the human resources system of production will be discussed. When *PSM1* was published in 2003, the highest risk group statistically for workplace injuries was young workers. With the shifting average age of the workforce, this has changed and will become a more prominent issue for organisations to manage as time ticks on.

REFERENCES

ABC News, 2014, Ultramarathon runner Turia Pitt, burnt during race in Kimberley, WA, reaches multi-million-dollar settlement. ABC News.

ABC News, 2019, Brazil's Brumadinho mining disaster will hurt Vale, but iron ore firms and Australia's economy are set to cash in. ABC News.

ABC News, 2000, Hungary to sue Australian company over cyanide spill › News in Science (ABC Science).

Australasian Mine Safety Journal, 2019, Hazard management. Australasian Mine Safety Journal (amsj.com.au).

American Society of Quality, 2024, 5S - what are the Five S's of lean? ASQ.

Bajwa, U., Gastaldo, D., Di Ruggiero, E., et al., 2018, The health of workers in the global gig economy. *Global Health*, 14, 124. https://doi.org/10.1186/s12992-018-0444-8

Burke, M., Driscoll, A., Heft-Neal, S., & Wara, M., 2021, The changing risk and burden of wildfire in the United States. PNAS.

Civil Aviation Authority UK, 2024, How does bowtie work. Civil Aviation Authority (caa. co.uk).

Deloitte, 2021, Practical guide to workforce ecosystems. Deloitte Insights.

Duong, C., 2023, Geotechnical expert says one in four international tailings dam failures are due to earthquakes, Earthquake research shaking up tailings dam safety. The National Tribune.

Forbes, 2020, The most valuable skill in difficult times is lateral thinking—here's how to do it (forbes.com).

Forbes, 2022, Here's what an aging workforce means for America's employers. Fortune.

Faure, M., Jing, L., & Wibisana, A.G., 2014–2015, Industrial accidents, natural disasters and act of God 43. Georgia Journal of International and Comparative Law 2014–2015 (hei-nonline.org).

Government of United Kingdom, 2020, New measures to improve building safety standards. GOV.UK (www.gov.uk).

Hansen, S., 2020, Does the COVID-19 outbreak constitute a force majeure event? A pandemic impact on construction contracts. *Journal of the Civil Engineering Forum*, 6, 201–214. https://doi.org/10.22146/jcef.54997.

Jansz, J., 2017, *International Encyclopedia of Public Health*, Second Edition, Editor in Chief, Stellar R. Quah, Elsevier, Amesterdam. https://www.sciencedirect.com/referencework/9780128037089/international-encyclopedia-of-public-health#book-description

Kaine, S., & Josserand, E., 2019, The organisation and experience of work in the gig economy. *Journal of Industrial Relations*, 61(4), 479–501. https://doi.org/10.1177/0022185619865480

Kamp, A., & Le Bansch, K., 1998, Integrating Management of Occupational Health and Safety and Environment: Participation, Prevention and Control. Paper Presented to the International Workshop: Policies for Occupational Health and Safety Management Systems and Workplace Change, Amsterdam, 21–24 September, p. 6.

Keary, P., Brooks, M., & Hill, I., 2002, Chapter 1: The principles and limitations of geophysical exploration methods. In *An Introduction to Geophysical Exploration*. Oxford: Blackwell Publishing, Kearyintro.pdf (blackwellpublishing.com).

Mercante, J.E., 2005–2006, Hurricanes and act of God: when the best defense is a good offense 18. University of San Francisco Maritime Law Journal 2005–2006 (heinonline.org).

MITSloan Management Review, 2021, Workforce ecosystems: a new strategic approach to the future of work (mit.edu).

NC State Poole College of Management, 2016, The bow-tie analysis: a multipurpose ERM tool. ERM - Enterprise Risk Management Initiative. NC State Poole College of Management (ncsu.edu).

News.com.au, 2011, Marathon fire: runners Turia Pitt and Kate Sanderson fight for their lives. news.com.au — Australia's leading news site.

Occupational Safety and Health Administration, 2024, Alphabetical listing of topics. Occupational Safety and Health Administration (osha.gov).

Perry, Lance S., 2010, The aging workforce: using ergonomics to improve workplace design. *Professional Safety*, 55, 22–28. The Aging Workforce: Using Ergonomics to Improve Workplace Design. Semantic Scholar.

Piciullo, L., Storrosten, E.B., Lui, Z, Nadim, F., & Lacasse, S., 2022, *A New Look at the Statistics of Tailings Dam Failures*. ScienceDirect.

Rezaei, A., & Caulfield B., 2021, Safety of autonomous vehicles: what are the insights from experienced industry professionals? *Transportation Research Part F: Traffic Psychology and Behaviour*, 81, 472–489. ISSN 1369-8478, https://doi.org/10.1016/j.trf.2021.07.005. https://www.sciencedirect.com/science/article/pii/S1369847821001649

Safe Work Australia, 2023, Glossary. Safe Work Australia.

Schwartz, A.A., 2020, Contracts and COVID-19 essay 73. Stanford Law Review Online 2020–2021 (heinonline.org).

Sherratt, F., Thallapureddy, S., Bhandari, S., Hansen, H., Harch, D., & Hallowell, M.R., 2023, The unintended consequences of no blame ideology for incident investigation in the US construction industry. *Safety Science*, 166, 106247. https://www.sciencedirect.com/science/article/pii/S0925753523001893

Stanford News, 2021, Lessons from Fukushima disaster 10 years later. Stanford News.

Swanson, J.A., 2013, The hand of God: limiting the impact of the force majeure clause in an oil and gas lease 89. North Dakota Law Review 2013 (heinonline.org).

Transurban Group, 2023, Connected and automated vehicle trials. Transurban Group.

U.S. Bureau of Labor Statistics, 2021, Number of people 75 and older in the labor force is expected to grow 96.5 percent by 2030: The Economics Daily: U.S. Bureau of Labor Statistics (bls.gov).

United States Environmental Protection Agency, 1991, Sick_building_factsheet.pdf (epa.gov).

US Fire Administration, 2022:

Story: High-Rise Fire Safety (fema.gov).

Protecting people who live or work in high-rises (fema.gov).

Weil, D., 2014, *The Fissured Workplace: Why Work Became So Bad for So Many and What Can Be Done to Improve It*. Cambridge, MA: Harvard University Press.

World Health Organization, 2017, Report & presentations of a joint symposium on the indoor environment & respiratory illness, including allergy, PCS_98.16.pdf (who.int).

6 Human Resources

A SYSTEMS APPROACH TO HUMAN RESOURCES RISKS

In *PSM1*, the term 'resourcefulness' was used to illustrate the overflow or benefit of building organisational capacity. This was tied in with the Entropy Model and as a result, 'resourcefulness' entailed the implementation of the four-step risk management strategy which, in turn, should result in better systems review, improved problem-solving and decision-making, effective leadership at all levels and systems of production raised towards optimal performance levels.

Since *PSM1*, there has been significant debate regarding whether humans are hazards or heroes in the OSH management field. In this discussion, it's worth expanding this to hindrance or hero when it comes to efficiency. Much of the research done in human factors has involved high-reliability organisations (HROs). The work of James Reason (2008) explored human contribution to the reliability and resilience of complex, well-defended systems. It was argued that humans may be considered hazards when taking the stance of unsafe acts implicated in many catastrophes. The human as a hero on the other hand results from adaptations and compensations that bring troubled systems back from the brink of disaster (Reason, 2008).

Other writings including Sandom (2007) highlighted that whilst human factors are often cited as the cause of hazards within safety-related systems, safety case studies at the time, often didn't refer to them. In practice, this has now matured in many organisations that include people-related risks such as fatigue, drug and alcohol abuse, bullying, violence, and other human hazards, in their risk register. This has to a significant extent been brought about by legislative requirements. The types of behavioural factors included in risk reviews tend to be issues that can be neatly packed based on statistical evidence. For example, it's a fact that fatigue impairs the worker's ability to perform safely. Driving under the influence of drugs or alcohol causes disfunction that compromises safety. Cognitive deficiencies however are much more difficult to categorise due to complexity. This is also obstructed by the broad scope of human factor analysis ranging from organisational culture to human error (Sandom, 2007). No single approach provides a clear understanding of the risks.

Recently, Bagley et al. (2023), wrote about hero or hazard by undertaking a systematic review of individual differences linked with reduced accident involvement and success during emergencies. The findings were that involvement in accidents was predominantly affected by cognitive ability, leadership, situational awareness, personality and risk perception. From an implementation perspective, the findings point to recruitment and training of frontline workers in HROs and other organisations to reduce risk and increase safety.

DOI: 10.1201/9781032701561-7

Unfortunately, when incident findings consistently give rise to the need for training and more training, this doesn't help operations to better understand HR risks, how to assess those risks or to devise strategies for better management.

What's useful however, is to continue the shift away from accident causation based on unsafe acts and conditions to a systems perspective. Human error needs to be contextualised within the entire system as an element of the HR system of production that has interfaces with technologies, processes and physical workplaces. There are often deficiencies at these interfaces that increase the probability of errors, for instance, poorly designed road intersections or inconsistencies in how levers and dials are operated on a control panel. The question is, has the system been 'fool-proofed' (without trying to define the characteristics of a 'fool')?

Historically, behaviour-based safety (BBS) programmes have been flouted as the panacea for human error, but these programmes are now being challenged and correctly assigned as one option within the overall risk management strategy. The systems approach acknowledges human fallibility and that errors will occur. The following quote although more than 20 years old continues to hold true:

> The situation in which the injury is actually caused by the worker, while existent, is extremely rare.
>
> The statement that 80 percent to 95 percent of accidents are caused by unsafe acts is wrong because the primary causes come from management systems and the facility.
>
> The great majority of actual causes of injuries is an interaction between the worker and the facility.
>
> This point of interaction is the working interface, which should be the subject of attention.
>
> Figuring out how the worker interacts with the system changes the focus of improvement from the worker to systems that enable safe behaviors.

(Manuele, 2000)

PSM1 concentrated on this working interface. According to the Entropy Model, HR are to be maintained at optimal performance, safety, and quality to reduce the probability of an accident or loss, which does present a strong case for investment in training and development. Unfortunately, most risk management training tends to be classroom-oriented and OSH-focused. There's a gap in connecting risk management competencies to production, safety and quality performance, as an integrated competency. In *PSM1* and this edition, capacity building involves learning as an ongoing process through participative activities and the day-to-day successes and failures of operations (as described in later chapters).

In *PSM1*, a principle was applied that employees are assets (not as rhetoric but as a matter of fact). The traditional approach to OSH management portrayed them as a liability – the primary source of unsafe acts and accidents in the workplace. In addition, accounting practices record labour as an expense and technology as an asset.

Fundamental accounting practices won't change but the treatment of HR as an asset is a necessary and important philosophic shift for organisations intending to build sustainable businesses for the future.

There has however been traction since 2000. Human capital is now considered the key to a successful Environmental, Social and Governance (ESG) Strategy (World Economic Forum, 2021). The trend includes the application of sustainability accounting which hopefully, will in the future, result in individuals accumulating quantifiable monetary asset value through experience and transferable competencies such as risk management skills. Perhaps, this will assist organisational decision-makers to take a longer-term, risk-based view when cost-cutting pressures are applied by the board or executive, and the easiest way to save money is to downsize workforce numbers. The questions should be, "How much have we got invested in this human asset?" "Can we afford to lose this human asset (especially to a competitor?)"

More generally, there remains a significant gap between corporations with established and evolving ESG Agendas and small to medium enterprises (SMEs) when it comes to perceptions of the asset worth of personnel. This is a broad and complex landscape. Traditional management styles are still prevalent in immature organisations. The dynamics can be characterised by power aligned with authority that, in the event of a safety incident or production loss, result in errors being detected and the 'delinquent' being pointed out (Pierce, 2000). Even in these organisations, blaming the worker has created a quandary because in most cases, managers and supervisors know that the worker doesn't choose to be injured, and they as supervisors don't intentionally set out to harm workers. Alignment on values continues to be talked about in management literature but achieving it is clouded with debate including the current emphasis on 'no blame culture' which is having some unconstructive consequences as discussed in the previous chapter. According to Topf (2000):

> Human behavior in the workplace is conditioned by a number of factors. It is not only governed by corporate imperatives but is conditioned by the employee's values. Thus success of the enterprise depends in large measure on the extent to which these two value systems – the corporate and the personal – are in harmony... In other words, to achieve a safe workplace, we need leadership that supports safety, health and environmental excellence as a reflection of values.

(Topf, 2000)

The crux of the ongoing complexity, philosophical debates and various versions of 'safety', such as Safety I, Safety II, Safety Differently and Synesis, is that human resources continue to be viewed through the psychosociological lens. What's needed to balance this is a risk-based approach to help reduce subjectivity. As discussed in Chapter 4 about technological risks, mathematical modelling using Shannon's entropy method is assisting with this. Ali et al. (2024) for instance, used Shannon's Entropy with other mathematical processes to assess failure modes and effects on the task of cutting core samples in a laboratory. Shannon's entropy is employed to determine a degree of disorder. The smaller the entropy value the less disorder in the process or systems. This is then used for criteria weight calculations. Conclusions included:

> This case study offered a unique approach for the HSE risk evaluation of different hazards (as failure modes) present in the core laboratory. The approach adopted here is quite flexible and reliable and addresses the limitations or drawbacks of previously widely known conventional techniques like the Risk Matrix based technique or FMEA in many ways...
>
> the risk outcomes obtained were extremely precise and provided a clear distinction between the risk level of different failure models, which is crucial for their comparative ranking.
>
> Though the scope of this study is limited to the risk ranking of HSE concerns, such as failure modes, of a particular experiment, however, the outcomes of this research work will certainly benefit organizations towards building sound HSE culture in Labs. Some of the benefits which are expected to achieve as a result of this study are:
>
> - Identification of the role of risk ranking in setting and revisions of policies, strategic goals and objectives and key performance indicators
> - Development and revision of safety plans, processes, procedures and guidelines...
> - Identification of an effective and sufficient training needs and professional development and growth programme of personnel...
> - Allocation of required human, financial, and physical resources for monitoring and mitigating the risk levels of hazards...
>
> *(Ali et al., 2024)*

What's particularly interesting about this approach is that it eliminates debate about whether humans make mistakes. Different failure modes include looking of machine parts during operation; short circuit due to heating/spark; loosening and striking of the core during operation; and contact of a body part with a moving blade. The method suggested is intended to support decision-makers with the result being 'maximum group utility' for the 'majority' and the 'minimum of individual regret' for the 'opponent' (Ali et al., 2024). The latter terminology isn't explained clearly in the article; however, it could be assumed that the work delivers optimal return on risk reduction (maximum utility) to address protection of workers (the majority), whilst also giving the decision-maker greatest confidence in their chosen actions (minimum regret).

Operations don't have the capability or resourcing to do this type of deep dive analysis of task risks, so a simpler approach is needed. The takeaways from this study are valuable.

Firstly, the maths agrees that as entropy rises so does the risk.

Secondly, strategic risk management should focus on human resources objectively as a system of production, not on the complex organism or colony.

Thirdly, human fallibility is a fact.

The risks are at the interfaces with other systems of production.

HUMAN RESOURCE RISK MANAGEMENT STRATEGIES

RESIDUAL RISKS

HR risks, like the other systems of production, have both residual and entropic risks. The latter is primarily due to capability gaps. The classification as a residual not an entropic risk is because of the long lead time to develop knowledge, skills and abilities (KSAs) and reach optimal productivity, safety and quality work performance. It's not possible for an individual or group to have a comprehensive set of competencies to deal with every situation that may arise during the execution of the work. As a result, from time to time, workers will use technology ineffectively, cause the workplace to degrade or deviate from standardised (presumably optimised) work practices. Incorrect judgements will be made from failing to appreciate the risks. In some cases, a flawed decision may result from not applying embedded competencies, for instance, knowingly speeding to make an appointment on time. Some decisions may be reasonable and rational, such as an emergency. Cognitive limitations mean that a worker or team may be unable to assess the possible consequences of an action prior to an undesirable event. It's only with hindsight that relevant information is revealed, and if known, may have led to a different course of action thereby preventing the loss in the first place.

HR residual risks are concurrently associated with limitations of the human body. Body builders are aware of this when they intentionally work muscle groups to the point of failure. Stating the obvious, humans don't have eyes in the back of their heads so the senses have a narrow range and therefore, capacity to detect and react to threats. The variability between individuals means one size doesn't fit all and matching the person to the job is an important and widely accepted mitigation.

In current risk management practices, HR risks aren't readily discussed in this way perhaps because of fears of objectifying workers. These concerns are ill-founded as there are numerous examples of organisations having to adapt to meet the specific needs of workers with physical and/or cognitive limitations, such as disability services. In other industries such as health care, the risk source may be due to the patient profile.

It's important to step away from associating risk management with discrimination. The objective is to protect people from harm and enable productive work, not to treat them unfairly or downplay moral and ethical principles.

New employees are a significant source of residual risk, and in some high-risk companies, they're identified with a green hardhat to differentiate them from experienced workers. New employees increase the level of residual risk by diluting the overall capability of the collective workforce. They have a higher probability of introducing entropy in the form of deviations from safe or optimal practice as they interact with other systems of production. The root cause is lack of workplace-specific competencies which take time to be acquired and will vary amongst individuals based on factors such as previous experience in similar work environments. Proactive organisations implement an orientation and assimilation programme to accelerate new incumbents' learning curves. They regard new starters as a 'hazard' during this period and manage the risk through supervision and on-the-job mentoring.

Historically, young workers had the highest incidence of fatalities and serious injuries, as reported in *PSM1*. For instance, the lost time incidence rate for males aged 15 to 24 in Western Australia in 1996/97 was 7.7 compared to other male workers at 5.9. This pattern was not evident for young female workers who had a rate of 2.2 compared to other female workers with a rate of 2.6 (Worksafe Western Australia 1994/5). Recent statistics from the UK indicate that the fatality rate by age group per 100,000 workers for 2018/19 to 2022/23 was 0.16 for 16- to 19-year-olds, and 0.28 for 20- to 24-year-olds, with figures progressively rising by age group to 0.75 for 60- to 64-year-olds, and an alarming 1.43 for workers over the age of 65 years (Health and Safety Executive, 2023).

For the purposes of comparison between the earlier Western Australian statistics and recent results in the UK, in the ten years from 2011–2012 to 2022–2021, workers aged 65 and over made up only 4% of the workforce, but 16% of work-related fatalities occurred amongst this group (Department of Commerce, 2000–2001). A study in 2022 showed that in the construction industry, exposure to chemical substances caused the greatest number of fatalities among older workers followed by falls and then vehicle incidents (Kamaree and Hasam, 2022). The results overall indicated that accumulated exposures are now evident, exacerbated by higher rates of older workers remaining in the labour force.

Another more recent study indicates mixed results when it comes to fatality and injury rates by age. Bravo et al. (2020) undertook a systematic literature review of fatal and non-fatal injuries between older and younger workers. Fifty percent of the results showed more fatalities amongst the older workers but the other 50% were equally divided. When it came to non-fatal injuries, the results were also complex with 49% showing no relationship between age and non-fatal injury rates. Increased age was found to be a positive, protective variable for injury prevention in 31% of studies. The remaining 19% showed that older workers were more vulnerable to serious injuries.

The commentary in *PSM1* identified young male workers as the highest risk group. Work-related and road traffic fatalities and injuries continue to indicate that this is true, so it remains fair to say that age and experience, and the associated learned competencies attained over time, do matter when comparing the residual risk of one age group against another. The Entropy Model indicates the need to manage these residual risks. Messaging and culture are important in this regard, for instance, by conveying safety norms to influence young workers through their interactions with supervisors and co-workers, so they more quickly come to understand expectations of safety-related conduct (Pek et al., 2017). Providing a safe learning environment, however, is the most important risk mitigation during this period.

Cognitive and physical limitations determine the residual risk of individuals and collective human resources (employees and contractors). HR risk management therefore begins with understanding the profile of the workforce. If there's a significant percentage of young or inexperienced workers, this may warrant the need for more supervisors, particularly in high-risk areas. It may be possible to assign each young worker to a mentor or to engage them in a more controlled workplace with direct supervision in the initial few weeks or months. This is also a sound strategy for preventing entropic risks resulting from competency gaps.

Certain job categories are physically demanding. Again there's a lead time to get workers job-ready, particularly if the work environment is challenging. For instance, exploration drilling offsiders are required to work outdoors in harsh conditions (40°C+) and undertake tasks such as shovelling and carrying sample bags filled with earth for up to 12 hours per day, often for 13 days in a row before getting a day of rest. These jobs tend to be filled by young males chasing the money and adventure, who have varying levels of physical fitness and little previous experience in the industry. Employers need to consider each candidate on a risk-basis with the input of medical and fitness assessments as a condition of employment. The gap is that once this basic check is completed, it becomes part of an onboarding 'tick and flick' with little follow up by the supervisor in the field who could be under pressure to get the new person working to full capacity as soon as possible.

Across high-risk industries, the missing risk management work is that managers and supervisors are seldom given straightforward risk-based information. HR residual risks need to be considered in the context of the residual risks of the other systems of production. Inexperienced workers shouldn't be left unsupervised or without a mentor. The risks of injury, damage, inefficiencies and other issues can be significant.

In contrast, well-managed HR residual risks can have the opposite impact enabling the business to continually improve. In practice, this is most often achieved by having a core group of experienced operational leaders. They provide direction amidst the turmoil until reliable, quality systems are in place for better personnel recruitment, selection, onboarding and workplace coaching. The worst decision a business owner or manager can take is to ignore the 'revolving door' of job incumbents as normal to the industry.

Accepting the status quo is a recipe for ongoing losses many of which have a financial impact, although these may be more indirect than direct costs. The astute risk professional faced with these challenging circumstances will do math totalling the cost of bringing each new incumbent onboard multiplied by the annual turnover rate of that job category, to equate the annual direct cost. Other indirect costs can be calculated from injuries, including loss of productive time of the worker, the supervisor and support staff, to put forward a case for better 'front-end' HR risk management embedded in the recruitment and assimilation process.

A dollar-based strategy is sometimes required to break the management attitude that leads to 'breaking' people.

ENTROPIC RISKS

The residual risks of the HR system of production are listed in Table 6.1 along with examples of risk management strategies. As explained earlier, these are derived from incomplete KSAs and the physical limitations of human factors. Many organisations implement these strategies but don't necessarily articulate the nature of the risk as clearly as provided in the table.

TABLE 6.1
HR Risk Management Strategies

Source of Risk	Residual Risk Management Strategy
Incomplete knowledge, skills and abilities (KSAs)	• Induction and task-specific training • Risk management training • Training prior to introduction of changes in processes, technology and/or physical environment • Effective communication systems • Coaching and mentoring
Physical limitations and fragility	• Modification of the physical environment, technology and processes to fit the worker • Matching the worker to the demands of the job • Provision of sick leave entitlements • Job rotation and job enrichment to share physically demanding tasks • Training in manual handling techniques • Procedures for the management of biohazards • Preventative health measures
	Entropic Risk Management Strategy
The new employee	• Value-based recruitment and selection procedures • Induction • Coaching and mentoring • Pre-operational training • Hazard monitoring of interaction of new employees with other system factors
Fatigue	• Reducing the demands imposed by the physical environment, technology and processes through modification and maintenance • Regular rest and meal breaks • Job rotation and/or enrichment strategies • Management of manual handling hazards • Monitoring of operator alertness by supervisor, co-workers and/or technological means • Training in self-monitoring and management strategies • Effective management of shift work rosters and practices • Flexible systems of work • Fatigue management training
Ill health (physical and mental)	• Multi-skilling and flexible systems of work • Monitoring of operator alertness by supervisor, co-workers and/or technological means • Training employees in self-evaluation of physical capacity • Healthy lifestyle programs • Employee Assistance Programs (EAP)
Stress	• Legally compliant and socially responsible operations - hazard minimisation • Planned and well-managed organisational change • Appropriate and adequate resourcing strategies • Stress management programs • Dispute resolution procedures • Employee Assistance Programs (EAP)

(Continued)

TABLE 6.1 (CONTINUED)

HR Risk Management Strategies

Source of Risk	Entropic Risk Management Strategy
Psychosociological risks including bullying, discrimination and violence	• Well communicated policies and procedures • Dispute resolution procedures • Psychometric testing for leadership positions • Whistleblower hotline • Procedures for dismissal where appropriate • Employee Assistance Programs (EAP)
Safety violations	• Organizational culture and practices which discourage violations and risk-taking • Communication of safety policy and duties of employees • Risk identification and management training • Balanced scorecard-based incentive programs • Disciplinary procedures • Procedures for dismissal where appropriate
Drug and alcohol abuse	• Substance Abuse Policy • Monitoring of worker capacity by supervisor and co-workers • Pre-employment drug/alcohol testing • Random testing • 'Moderation' culture • Healthy lifestyle programs • Employee Assistance Programs (EAP)

As shown, there are also numerous sources of HR entropic risk. The enterprise is particularly exposed when bringing new personnel onboard. This risk becomes pronounced if the business suffers from high turnover such as those relying on seasonal or migrant workers. Traditionally, turnover has been considered a metric for the HR (People and Culture) Department to measure and develop programmes for reduction; however, Operational and OSH Departments should also be involved in strategy development because turnover has serious implications for safety, production and quality from a risk management perspective.

Regardless of competencies or age, workers who are in a degraded state can have an accident or trigger a loss. A highly skilled worker suffering from fatigue has less capacity to perform the work efficiently, safely and to the required quality standard. Although skilled workers have a lower residual risk, the susceptibility to deterioration is similar if not the same as less skilled colleagues. The difference may be in who recognises the signs and takes timely corrective action by reporting their condition or taking a rest break.

Fatigue is an entropic risk that can rise rapidly. There's extensive literature on the causes of fatigue and in summary, these can be organisational and/or individual factors. Due to the latter, education is an important element in the fatigue management programme. HR entropic risks vary according to the interface with other systems of production. For instance, demanding physical environments take a toll on both body and mind. The technology involved in undertaking the work also has a significant impact; for instance, compare riding a motorbike to a car with other variables consistent – the same speed, road, other vehicles and site conditions. OSH regulators provide numerous resources online to enable the effective management of fatigue (NIOSH, 2023; Health and Safety Executive, 2023, Safe Work Australia, 2023). Whilst fatigue management programmes tend to fall within the remit of the OSH function, the impact on production performance means that operations and HR departments should have input to these programmes. Many of the recommendations proposed by regulators include hours of work, design of shifts and rosters, breaks and other matters that are usually incorporated into the employment / engagement contract and work planning.

The fatigue management education piece is mandated in some industries such as commercial drivers. The gap from a broader industry perspective is that organisations (or perhaps the education system) should move towards risk management as a life skill. This would encourage people entering the workforce to take ownership of managing risk as an ongoing personal competency. This in the longer term should create a more open learning environment centred on risk-based conversations within organisations.

Some strategies for fatigue management are listed in Table 6.1 which includes mental fatigue and physical fatigue, for instance, the latter resulting from manual handling. This needs a renewed effort given the issues that are arising with an aging workforce. Other types of HR entropic risk shown in Table 6.1 include ill-health and stress, both of which are extensively covered in resources readily available online. *PSM1* provided a detailed description with legislative references applicable at the time.

What's worth noting here is that mental ill-health and stress, when suffered by a person with significant influence on a team or within the organisation, warrants a concerted effort by the enterprise and by that person's support network. This is because of the flow on effect.

A leader whose performance is degrading can create a ripple effect that brings the rest of the team down professionally and, in some cases, personally.

This will be illustrated conceptually using the Entropy Model in terms of organisational capacity in Chapter 8. Suffice to say for the moment, that leadership can degrade causing a rising organisational entropic risk.

My Thi Diem Ta et al. (2022) studied the importance of leadership styles on safety, using the traditional term 'safety leadership'. They reviewed nine styles that have been frequently used in the development and validation of safety leadership theories. This may be useful from an academic perspective. In the workplace from a risk perspective, it's the strategic and operational decisions leaders make and how they go about engaging others in the process that matters. In *PSM1* and this edition, there's only 'leadership' – not 'safety leadership'. If a manager is ill, stressed, has relationship issues or other factors, which impair their decision-making abilities, then production, safety and quality (including the quality of the work experience for others) are likely to suffer. Ill-health, stress and psychosociological risks are identified in Table 6.1 as entropic risks.

The better approach is to put aside 'safety leadership' and focus only on 'leadership' (including when reading the 'safety' literature), to understand the risks of degradation and the triggers for corrective action. With this broader lens, there are more opportunities to learn what works. For instance, Sadiq (2020) found five factors by which leaders create a workplace culture for injury prevention (which can be expanded here to 'loss prevention'). Broadened beyond safety, these were empowering employees; promoting engagement; creating a learning organisation and culture; having an effective process-based management system; and modelling leadership behaviour. It can be inferred that if any of these falters then senior management need to take corrective action, in a manner that is appropriate on a case-by-case basis but with sound risk-based guidelines and a multidisciplinary approach.

What is at stake if an organisation chooses not to act when a leader's performance declines? This is the type of question that has raised the use of the term 'undiscussables' in more recent years and it's particularly relevant when it comes to executive stress and performance (Rook et al., 2016). Another study based on academic health centres recommended that enterprises increase capacity to prevent organisational silence. These include developing leaders to counteract silence; developing skills in raising difficult issues with those in positions of power and training mentors to coach others to raise difficult conversations (Dankoski et al., 2014). According to Sterling (2020), corporate disasters such as Enron and Theranos were preceded by undiscussables in the upper echelons of the businesses. This quashed challenges and created a culture of fear and denial. Stifled voices caused interactions to be politically motivated and oppressed the pursuit of truth. The lesson to be learned, according to Sterling, is that leaders must learn to prefer followers who aren't simply obedient and who make them think.

Just as 'black swans' are unknown unknowns in technologies and the physical environment, 'undiscussables' create blind spots to leadership degradation that can result in loss of capacity to deliver on organisational objectives.

This discussion provides food for thought. From a risk management perspective, operations managers and other key decision-makers should be made aware of these HR entropic risks in terms of their teams' performance but also as a barometer for self-reflection by those in positions of authority. (This will be discussed in greater detail in Chapters 8 and 9 covering leadership and organisational capacity, respectively.)

A further source of entropic risk in the table is psychosocial risks including bullying, discrimination, violence, safety violations, and drug and alcohol abuse. It's in relation to these risks that there needs to be clear accountabilities; specifically, differentiation between what the organisation controls and influences compared to behavioural choices made by individuals. Psychosocial risks and substance abuse are acknowledged as far as mitigations are concerned because of legislative requirements and guidance material available on regulators' websites. (These topics won't be covered in this edition of PSM as there are extensive resources available in the public domain.) 'Safety Violations' on the other hand are a grey area that the OSH profession and line managers continue to struggle with. Fortunately, the impact of production pressure on operational behaviours has become a focus area for researchers so some recent studies can be used to clarify what constitutes a 'safety violation' versus an organisational factor.

PRODUCTION PRESSURE AND SAFETY VIOLATIONS

Table 6.1 lists some strategies for addressing safety violations and these primarily follow traditional OSH and HR practices. Safety violations and drug and alcohol abuse can be put into a similar category because these both are driven to some extent by individual factors, although violations aren't always that simple.

It would be easy to apply the criteria of 'intent' to acts and then classify them as violations. For instance, the worker intentionally commits an 'unsafe act' which they know doesn't comply with a documented procedure, for instance, leaving the engine running and exiting a vehicle is considered a breach of safety on most Australian mine sites but is it this simple? What if the supervisor does this on a regular basis? What if the worker has been observed doing this numerous times without being advised of the correct parking protocols? What if on all other occasions, no incident occurred but this time, the vehicle has rolled and crashed into another vehicle narrowly missing a pedestrian, so the behaviour has been 'caught'?

In practice, a 'violation' is not simply intent.

It's also a matter of whether the behaviour is non-compliant to required practice – procedural (written), the norms of the workgroup (in practice) and the organisational culture (expectations).

Having a procedure, getting workers to read it and sign-off that they've read it, isn't enough. Unfortunately, this is still common practice, and managers often assume that this will protect the business and themselves from accountability if a safety incident occurs. In immature organisations, obtaining workers' signatures after reading documents lays the path open for blame during incident investigations. The OSH professional who leads the enquiry must carefully navigate this type of scenario with the knowledge that numerous court cases, including *Barcock v.* Brighton Corporation 1949 (described earlier), have established precedent that employers/PCBUs can't simply pass the risk to the worker after providing basic information and training. Workers must be actively engaged as part of the work in identifying risks and deciding how these are managed. Monitoring regimes must also be in place that test the effectiveness of risk controls with a feedback loop that triggers action if controls are inadequate.

Many commentators describe active engagement of workers in risk management as central to building a strong 'safety culture'. As mentioned earlier in Chapter 2, various disciplines will have their own framework for 'culture' such as environmental protection, quality, or lean culture. This alone can create conflict and non-alignment on goals and values because of intrinsic tensions. Safety that becomes bureaucratic inherently becomes inefficient as suggested by Hollnagel's Efficiency Thoroughness Trade-Off. Operations often bear the brunt of this dilemma, which in turn can affect behavioural choices including procedural deviations to get work done quickly, even if this means 'non-compliance' to the system.

Historically, various accident causation models were used to describe how people make behavioural choices in the presence of risk. These can be helpful to better understand safety violations. Risk Homeostasis Theory (Wilde, 1998) is a good example (although not without its critics – Trimpop, 1996). In summary, people have an expected utility of action alternatives which means they expect some benefit from various choices available to them. This affects their target level of risk they consider 'acceptable'. This may be higher than that of the organisation. Individual perceptive skills are used to assess the level of risk of a particular activity. The target and perceived levels are compared which leads to adjusted action or choice of behaviour.

In practice, the organisation needs to, firstly, clearly define what isn't acceptable behaviour, for instance, using golden rules or strict high-risk procedures such as isolations, lockouts, and permits to work (This is where traditional approaches to safety continue to have merit). As it's impractical to have rules for everything, other risk-based decisions must be influenced to reduce the expected benefits or increase the expected costs of risky alternatives. A cost, for instance, would be disciplinary action if caught. On the other hand, for safe behaviour alternatives, the expected benefits can be increased and expected costs decreased (Wilde, 1998). The benefit might include positive feedback from the supervisor.

The discussions so far are alluding to the complexity of identifying a safety violation as a root cause of a safety-related incident. Some events are obvious such as a workplace argument that results in a worker breaking another worker's nose in the presence of witnesses, but most incidents aren't so straightforward. Consider the following scenario.

A worker is approaching the end of the three-week roster and in a couple of hours is due to fly home for a week. The cultural norms of the work team include preparing the plant and equipment for the next crew who will fly-in for their roster. The worker decides to clean down the truck which has a crane mounted on the back. The crane hoist is raised so that cleaning can be done thoroughly. This is extended in the air approximately eight metres. The worker receives a phone call and goes into the workshop. After this, the decision is made to refuel the truck, so the worker goes directly to the cab of the truck and drives out of the yard onto public roads. The extended crane brings down several overhead powerlines before the worker receives an alert from a colleague and stops the truck.

How easy would it be to jump to the conclusion that this was a safety violation? Perhaps, the worker could be classified as 'stupid' and the question asked, 'How often has this person been involved in previous incidents?' to determine if the person is a 'risk-taker' and 'liability' to the organisation. What impact would additional information have, for instance, that the worker is a supervisor with ten years' experience with the company? Would this change perceptions of guilt? Such matters which are person specific, introduce a cognitive bias which means there's drift into subjectivity.

What if the investigation revealed that the truck was parked close to the open large main entry of the workshop and that this proximity, meant that the arm of the crane was obscured from the supervisor's peripheral vision by the wall of the building. As a result, the supervisor didn't see the extended crane arm, forgot that it was still raised and went straight to the cab then drove off. No warning lights or alarms went off due to damaged wiring inside the console. This wasn't visible to the worker.

In the scenario given, the supervisor was trying to do the right thing by refuelling the truck before going on leave so that it would be ready for the next crew. The immediate contributing cause of the incident was that the alarm system that ordinarily warned of the raised crane arm wasn't functioning. Other gaps, for instance, in maintenance, inspection and testing, were derived from this. The starting point for the investigation was technological not human. The last area of investigation was person-related, for instance, as the incident occurred at the end of a three-week roster, was the worker suffering from fatigue? Would fatigue have changed the outcome? Probably not. When the worker's psychological state after an incident is considered, the systems approach rather than the person-centric approach, is much fairer. There's an opportunity to build trust in the process, rather than making them feel that they're under interrogation.

From the outset of incident investigation, assume that the person/s involved were trying to do the right thing for the right reason. Ask how this could happen? Questions what occurred at the interface between HR and the other systems of production. Analyse person-specific residual and entropic risks last.

This methodology is needed to reduce cognitive bias of the incident investigator / investigation team and others who may seek to have input or influence the process or outcomes. Inherent bias in logical reasoning was a basis of research done by Kahneman and Tversky (1972, 1974). A recent study presents cognitive illusions and explains who can solve them and why (Bruckmaier et al., 2021). In their experiment, it was found that the best predictor of ability to solve the brain teasers was mathematical competence followed by intelligence. Correct answers were more likely when there was a facilitator to translate information into an accessible form, otherwise, participants tended to fall into mental traps built into the brain teasers and get the answer wrong.

The lesson for organisations is to ensure that major incident investigations are led by senior personnel who preferably aren't directly involved in the operation, perhaps from another site, who are independent of local organisational cultural issues and biases. The incident investigation system should be carefully developed to minimise or prevent prejudice, for instance, some systems still categorise immediate and root causes into 'unsafe acts' and 'unsafe conditions', which continue to point at worker behaviour and incompetence (or at best, training gaps).

It should be borne in mind that many incidents could have occurred to someone else, not specifically to the person/s directly involved in a particular event. Prevention is about recurrence to protect anyone who could be exposed to the same or similar circumstances in the future. The other concern with starting with person-specific questions is that it sends unhealthy cultural messages that erode trust. (The work of Sidney Dekker including the recent publication of "Stop Blaming: Create a restorative just culture" provides deeper thought on this topic.)

It's also important to understand the context in which behaviours occur. In Chapter 2, the example was given of underground miners falsifying the pre-start checks on their vehicles and ignoring faulty park brakes. This was done to ensure that vehicles were available, not sent to the workshop for repair, so that they could push production, which would be rewarded through their next fortnightly bonus. This type of messaging drives the collective and individual target levels of risk upwards. Any incidents in this organisation, even if there were 'safety violations', would be difficult to justify in an environment where production pressure is the over-riding operational driver.

Evidence of production pressure and risk-taking behaviour in industry has been confirmed, for instance, by MOSHAB (1998, 1999). More recent studies include Hashemian and Triantis (2023) and Oswald et al. (2019, 2020). For instance, Oswald et al. (2019) did a study of the construction industry and identified that subcontractors were more likely to be paid by 'piecework' rather than 'by hour'. This creates inherent production pressure by incentivising them to work as fast as possible. A job completed quickly with safety shortcuts attracts the same remuneration as a job done in a planned, risk-managed manner that takes longer. The researchers identified that there were informal production management processes in place on construction sites that increased safety risks, and this was common practice. The Alignment Fallacy discussed in Chapter 2 explained this in the broader context that having a safety policy (even 'Safety First') doesn't ensure that policy becomes an operational reality.

The evidence of production pressure indicates the need to be wary of shortcuts in incident investigation that could result in personnel wrongly being accused of safety violations because they rushed the job.

Intentional violations of safety within effectively planned and well-controlled work environments, may occur and these can be addressed through the integration of OSH and HR management processes, for instance, coaching, supervision and/or disciplinary actions if appropriate. The primary means of preventing undesirable behaviours, however, must be, firstly, to remove areas of non-alignment such as production-centric incentive payments for contractors and sub-contractors, and secondly, to develop an organisational culture that discourages risk-taking. This is more difficult for short- to medium-tenure contracts than in steady state operation.

Contractor management and prevention of entropic risk is complex. Ideally, Operational Managers should regard contractors and subcontractors the same as permanent employees for ethical reasons but there are several obstacles to this. The first is the lack of long-term relationship, so where possible, service arrangements should be set up to integrate these suppliers. The second obstacle is procurement processes are cost-driven with work often going to the lowest priced tender. The third issue is that often these services are called upon with a sense of urgency either because they're specialist skills that aren't available in-house or because of absences or under-staffing, internal skills aren't available on the day. The contractor therefore arrives at the site and the client communicates pressure-driven messaging about the need for the work to be completed as soon as possible.

From a legislative perspective, the client has a safety duty to employees, contractors, and subcontractors but operationally, the process of engaging these workers can itself be riddled with shortcuts. These issues point to the potential for an organisation to have a broader operational culture which may be risk-based, but still have pockets of sub-culture that are production- or schedule-centric. Again, context becomes a factor in assessing an 'unsafe act' as it may be a safety violation or a symptom of systems or practices that have embedded cultural defects. The relationship between behaviours and culture will be discussed in Chapter 9 in which the two alternatives of risk-taking versus risk-managing will be presented using the Risk Behaviour Model (Van der Stap, 2008).

In the meantime, it's important to consider what constitutes a 'safety violation'? When an intentional unsafe act is committed by an individual despite the organisation having done their due diligence through the recruitment and selection process, providing necessary training and supervision, within the context of a risk-based culture, then it may be a violation.

For a safety violation to have occurred it must be preceded by individual behavioural risk-taking as the root cause, without extraneous influencing factors that are within the control of the organisation's systems or practices.

A warning needs to be provided as accountability remains fundamental to the decision process.

An individual or group in collaboration mustn't be able to blame the 'system' or the supervisor or production-pressure, as a means of avoiding accountability for a violation that has been identified based on objective evidence.

This needs to be borne in mind not just from an OSH perspective but also violations that cause other losses, for instance, to production performance, damage to equipment, environmental harm or quality defects. HR risk management must consider that whilst most people are trying to do the right thing for the right reasons, there may be some who aren't conscientious, ethical or willing to change. A systems-driven, risk-based loss investigation process needs to be in place to address sources of risk within the total system, which includes intentional damage or sabotage of that system. In Chapter 5, there was discussion about the 'No Blame Culture' leading to more incidents being ascribed to 'Acts of God'. This has been counter-productive for the OSH profession and safety performance, especially in the construction industry in the USA. What mustn't happen is that 'production pressure' becomes an avenue for avoidance of accountability.

The final section of this book will raise some of the opportunities for better HR risk management that are emerging from the Fourth Industrial Revolution. This will tie back to the discussions about the technological and workplace advances of previous chapters.

HR RISKS IN THE FOURTH INDUSTRIAL REVOLUTION

Some of the key developments of the Fourth Industrial Revolution were discussed in Chapters 4 and 5 about technology and the physical environment, respectively. Some other advancements have become more personal with widespread cultural acceptance, such as wearable technologies and mobile phone capabilities.

An interesting field from an OSH perspective is the use of exoskeletons to physically support workers or prevent musculoskeletal disorders. These reduce load whilst allowing the user to move freely and safely (IOSH, 2023). The uptake in aged care is an expected progression for the industry from simple supportive devices, but potential benefits in the workplace appear to be untested to-date (European Agency for Safety and Health at Work, 2019). Some concerns are being raised.

> In this context, industrial workplaces, furniture delivery services, emergency services and hospitals are of interest. However, it is important to recognise that ergonomic design namely in stationary workplaces. As long as technical or organisational measures provide possibilities for improving ergonomic design, the use of exoskeletons should not be preferred (Schick, 2018). Nevertheless, it seems that focusing on exoskeletons that increase the performance of workers might be of greater interest than focusing on the human-centred design of workplaces (Baltrusch et al., 2018).

> *(European Agency for Safety and Health at Work, 2019)*

The use of exoskeletons needs further research to determine the advantages and disadvantages of using them. These may have negative physiological effects from additional weight on the cardiovascular system. User acceptance will also play a major role along with potential psychosocial effects if workers feel inferior using them because of stigmatisation. Biomechanical aspects need to be considered as exoskeletons may support very specific movements and muscle activities and thereby influence the wearer's physical behaviour, which might not be an ideal fit for the range of work that needs to be performed.

The integration of wearables into OSH management systems appears to be on the rise. Various E-tool seminars are now available (European Agency for Safety and Health at Work, 2022) covering the challenges and benefits of their use in risk assessment. Smart digital systems are proposed to provide for safer and healthier workers (European Agency for Safety and Health at Work, 2023a, 2023b).

The overall success of these systems will rely on effective communication of data, standardisation, and inclusive implementation processes but worker privacy and data protection issues will also be key priorities (European Agency for Safety and Health at Work, 2023 a, 2023b) Large employer organisations are leading the way in Europe with most establishments having 250 or more employees. According to OSHA, current OSH legislation doesn't address the implications of technical change in OSH monitoring, making it a grey area. Those countries such as Australia which have Roben's style enabling legislation are probably better able to capture technological change within the OSH duty holders' accountabilities. Hesitation and lack of buy-in among workers and their unions is forecast to be an obstacle. Concerns are often centred on limited evidence of benefits and awareness coupled with fears of potentially hidden purposes of monitoring. Trade Unions are questioning whether these technologies will exposure workers to productivity pressures to the detriment of health and wellbeing (European Agency for Safety and Health at Work, 2023 a, 2023b).

Overall, the discussions about the Fourth Industrial Revolution in each chapter have flagged the need for organisational leaders to monitor developments for trends and opportunities. Professionals in strategic and operational risk-based disciplines, such as engineering, OSH, environmental and quality management, should maintain an open mind to the risks and opportunities that lay ahead.

SUMMARY

This has been the final chapter examining the residual and entropic risks associated with systems of production. There are extensive topics that could have been covered for HR risks that span multiple fields such as ergonomics, human operational performance, behavioural- based safety etc.

The aim of this chapter has been to describe the fundamental residual risks associated with cognitive and physical limitations of human beings and the impact these have on capacity to perform work safely, efficiently and as per required quality standards. The discussion highlighted human fallibility as a fact. Entropic risks were also explored. These fell into two main sub-categories – those that the organisation has primary accountability for and those that are largely dependent on individual decision-making factors. The residual and entropic risk management strategies were presented in Table 6.1.

Some of the more contentious issues associated with people management were raised. These included moving beyond the human as hero or hazard from an OSH and operational perspective. A systems approach is needed to contextualise HR risks including 'safety violations' to avoid unfair blame without glossing over accountability for failures. There are risks with adopting a 'No Blame Culture' that can be counter-productive to individual and organisational learning. There are no shortcuts to having clear accountabilities and responsibilities.

The OSH profession has been strongly influenced by the psychosociological lens of those considered to be thought leaders of the OSH discipline. Operationally, it has created confusion and blurred the sound, 'traditional' approach of incident investigation which must be an objective, technical process. Strategically, when aiming to use incident investigation as part of a healthy organisational culture, it's important to consider how the system failed the person (as a starting point) rather than how the person/ injured worker failed the system. The aim of a risk-based approach is not to objectify people (dehumanisation) but to create sufficient distance between the investigators and the personnel involved in the incident, to enable trust in the process. In immature organisations in particular, this approach can counter shortcut investigations solved by disciplining the person/s directly involved. The OSH profession should be cautious not to discard tried and tested practices.

Other disciplines are helping to make HR risk assessment more objective. This includes mathematics such as Shannon's Entropy Approach which looks for variations from the norm in human behaviours under a test condition. The method confirms rising entropy as a risk. The input of mathematics and big data analytics will likely make human performance analysis increasingly objective despite some societal concerns, which are likely to be addressed concurrently.

Since *PSM1* was published 20 years ago, there have been changes to the workforce HR residual risk profile in Western countries. This has resulted from the aging workforce. In 2003, the highest risk group were young workers primarily due to more risk tolerance and lack of experience. Currently and increasingly into the future, older workers are incurring injuries and dormant health issues are becoming evident. Risk management planning for this aging workforce needs to be done by organisations if injury rates are to be curtailed and occupational diseases detected early for treatment interventions. This was a risk that in the past was an 'undiscussable' as it raised concerns about discrimination. Times change.

One of the more contentious risks discussed in this chapter was the 'undiscussable' topic of leadership degradation. Personnel in influential positions can suffer from declining job performance for several reasons; however, interventions are often cautious and slow, resulting in risks to the performance of team members and subordinates. A more objective approach is needed to identify these HR risks and provide support. Traditionally, this was the realm of the HR Department, but OSH legislation has changed, and psychosocial risks are now considered an organisational 'hazard'. HR and OSH Departments must collaboratively manage this risk but where the person affected has operational accountabilities, the impact of their job performance on production/ schedule/ core business, must be part of the assessment process to prevent losses due to unreliable decision-making. What is at stake is not just safety or morale but other losses that affect the bottom line.

One of the key areas that a leader can fail is poor planning which results in production pressure cascading to the workforce. In recent years, production pressure is being recognised as a human and organisational risk that needs to be managed. Much more work needs to be done in this area in the research field and operational practice.

The chapter closed with the impact of the Fourth Industrial Revolution on HR risks. Personalised technologies are enabling data feedback to better understand human performance, for instance, using wearables. Exoskeletons are proposed as a future solution to reduce manual handling injuries and to facilitate safe movement for persons suffering from impairment. The cost–benefit analysis of these innovations is yet to be done for operations. In the bigger picture, the opportunities appear to be extensive and the need to better understand HR interfaces with other systems of production will be an important part of managing these changes.

In the next chapter, the residual and entropic risks of the four systems of production will be combined to present an alternative way of considering semi-quantitative risk assessment. This will be an update of Chapter 7 of PSM1.

REFERENCES

Ali, S.I., Lalji, S.M., Haider, S.A., Haneef, J., Syed, A., Husain, N., Yahya, A., Rashid, Z., & Arfeen, Z.A., 2024, Risk prioritization in a core preparation experiment using fuzzy VIKOR integrated with Shannon entropy method. *Ain Shams Engineering Journal*, 15(2), 102421, ISSN 2090-4479, https://doi.org/10.1016/j.asej.2023.102421

Bagley, L., Boag-Hodgson, C., & Stainer, M., 2023, Hero or hazard: a systematic review of individual differences linked with reduced accident involvement and influencing success during emergencies. *Heliyon*. https://doi.org/10.1016/j.heliyon.2023.e15006

Baltrusch, S.J., van Dieën, J.H., van Bennekom, C.A.M., & Houdijk, H., 2018, The effect of a passive trunk exoskeleton on functional performance in healthy individuals. *Applied Ergonomics*, 72, 94–106.

Bravo, G., Viviani, C., Lavallière, M., Arezes, P., Martínez, M., Dianat, I., Bragança, S., & Castellucci, H., 2020, Do older workers suffer more workplace injuries? A systematic review. *International Journal of Occupational Safety and Ergonomics*, 28(1):398–427. https://doi.org/10.1080/10803548.2020.1763609. Epub 2020 Jul 15. PMID: 32496932.

Bruckmaier, G., Krauss, S., Binder, K., Hilbert, S., Brunner, M., 2021, Tversky and Kahneman's Cognitive Illusions: Who Can Solve Them, and Why? *Front Psychol.* 2021 Apr 12. https://doi.org/10.3389/fpsyg.2021.584689. PMID: 33912097; PMCID: PMC8075297.

Dankoski et al., 2014, Discussing the undiscussable with the powerful: why and how... : academic medicine (lww.com).

Dekker, 2023, *Stop Blaming: Create a Restorative just Culture*. Dekker, Sidney, eBook - Amazon.com.

Department of Commerce, 2000–2001, Work-related traumatic injury fatalities in Western Australia 2011–12 to 2020–21p (commerce.wa.gov.au).

European Agency for Safety and Health at Work, 2019, The impact of using exoskeletons on occupational safety and health. Safety and health at work EU-OSHA (europa.eu).

European Agency for Safety and Health at Work, 2022, E-Tools seminar 2022- Wearable technologies for risk assessment. Safety and health at work EU-OSHA (europa.eu).

European Agency for Safety and Health at Work, 2023a, Smart-digital-monitoring-systems-Optimising-the-uptake_en.pdf (europa.eu).

European Agency for Safety and Health at Work, 2023b, New technologies for safer and healthier workers: the potential of smart digital systems for OSH. Safety and health at work EU-OSHA (europa.eu).

Hashemian and Triantis, 2023, *Production Pressure and its Relationship to Safety: A Systematic Review and Future Directions*. ScienceDirect.

Health and Safety Executive (UK), 2023, Work-related fatal injuries in Great Britain, 2023 (hse.gov.uk).

Health and Safety Executive (UK), 2023, Workplace fatigue management (hse.gov.uk). Institution of Occupational Safety and Health, 2023, Exoskeletons in the workplace. IOSH.

Kamaree, Imriyas and Abid Hasam, 2022, Occupational health and safety implications of an aging workforce in the Australian construction industry. *Journal of Construction Engineering and Management*, 148(10) (ascelibrary.org).

Kahneman, Daniel & Amos Tversky, 1972, *Subjective Probability: A Judgment of Representativeness*. ScienceDirect.

Kahneman, Daniel & Amos Tversky, 1974, Judgment under uncertainty: heristics and biases. *Science*, 185(4157), 1124–1131. TverskyKahneman1974.pdf (ubc.ca).

Law Insider, 2023, Barcock v Brighton Corporation [1949] KB 339, The claimant further relied upon the sample clauses. Law Insider.

Manuele, F.A., 2000, Behavioral Safety: Looking Beyond the Worker (Changing Theories of Facility and Worker at Risk Interaction). Occupational Hazards Oct 2000, Penton Media, Inc.

Mines Occupational Safety and Health Advisory Board, 1998, Risk Taking Behavior in the Western Australian Underground Mining Sector. Risk Taking Behavior Working Party Report and Recommendations.

Mines Occupational Safety and Health Advisory Board, 1999, A Review of Incentive-Based Remuneration Schemes in the Western Australian Mining Industry. Report and Recommendations from MOSHAB Incentive-Based Remuneration Working Party.

My Thi Diem Ta, Tae-eun Kim, & Anne Haugen Gausdal, 2022, Leadership styles and safety performance in high-risk industries: a systematic review. *Safety and Reliability*, 41(1), 10–44. https://doi.org/10.1080/09617353.2022.2035627

National Institute of Occupational Safety and Health, 2023, Work and fatigue: resources. NIOSH. CDC.

Oswald, D., Sherratt, F., & Smith, S, 2019, Managing production pressures through dangerous informality: a case study. *Engineering, Construction and Architectural Management*, 36(11), Emerald Insight.

Oswald, D., Ahiaga-Dagbui, D.D., Sherratt, F., & Smith, S.D., 2020, An industry structured for unsafety? An exploration of the cost-safety conundrum in construction project delivery. *Safety Science*, 122, 104535.

Pek, S., Turner, N., Tucker, S., Kelloway, E.K., & Morrish, J., 2017, Injunctive safety norms, young worker risk-taking behaviors, and workplace injuries. *Accident Analysis & Prevention*, 106, 202–210. ISSN 0001-4575, https://doi.org/10.1016/j.aap.2017.06.007

Pierce, F.D., 2000, 'Safety in the Emerging Leadership Paradigm', *Occupational Hazards* June 2000, Penton Media, Inc.

Reason, J., 2008, The human contribution. Unsafe Acts, Accidents and Heroic Recoveries (taylorfrancis.com).

Rook, C., Hellwig, T., Florent-Treacy, E., & Kets de Vries, M.F.R., 2016, *Stress in Executives: Discussing the 'Undiscussable'*. Kets de Vries: SSRN.

Sadiq, A., 2020, Influence of leadership practices on organizational safety performance (researchgate.net).

Safe Work Australia, 2023, Fatigue – resources. Safe Work Australia.

Sandom, C., 2007, Success and Failure: Human as Hero – Human as Hazard. 12th Australian Conference on Safety Related Programmable Systems, Adelaide.

Schick, R. (2018). Einsatz von Exoskeletten in der Arbeitswelt. Zentralblatt für Arbeitsmedizin, *Arbeitsschutz und Ergonomie*, 68 (5), 266–269.

Sterling, J., 2020, How leadership teams can face and fix their "undiscussable" dysfunctions. Emerald Insight.

Topf, M.D., 2000, Including Leadership in the Safety Process. *Occupational Hazards*. March 2000, Penton Media, Inc.

Trimpop, R.M., 1996, *Risk Homeostasis Theory: Problems of the Past and Promises for the Future*. ScienceDirect.

Van der Stap, T., 2008, Overcome the conflict between safety & production with risk management & behavioral safety principles, *RM Insight, American Society of Safety Professionals' Journal*, 8(2). pp. 12–16.

Wilde, G.J.S., 1998, Risk homeostasis theory: an overview (researchgate.net).

World Economic Forum, 2021, Human capital is the key to a successful ESG strategy. World Economic Forum (weforum.org).

Worksafe Western Australia, 1994/5, *State of the Work Environment No. 26: Injuries and Diseases in Young Workers*, Western Australia.

7 Risk Quantification and Management Strategy

THE ALARP ASSUMPTION REVISITED

Figure 3.4 in Chapter 3 illustrated the ALARP Assumption. It showed that as the operation and its risk management system mature, there is the danger it will move into an 'Operations: Low Value Thoroughness Phase'. Actions are taken because 'something' must be done, and this may add attention to risk reduction with diminishing returns. Other risks with unforeseen or unacknowledged consequences may arise such as production pressure as workers become confused by and bogged down with bureaucratic process. The excessive use of rules may have negative impacts on the culture, for instance, if workers perceive that they're in a 'parent–child' relationship with 'the system'. For this reason, the actual risk level into this low value phase was shown as variable and unreliable, and different from the ALARP level that management assumed had been achieved.

In this chapter, an alternative approach to risk assessment is presented based on the concepts of residual risk, entropic risk and the ALARP Assumption. The proposition is:

If the ALARP Assumption is correct, adding more lower order controls doesn't reduce risk. This means current risk assessment processes are deficient. A new approach that combines residual and entropic risks of an activity, and more broadly, 'The Work', needs to be considered.

The method proposed in this section is worth testing empirically. It serves several purposes for the risk assessment process and incident/loss investigation. The first purpose is to focus on risk and the potential for loss across multiple disciplines, thereby, implementing the rationale of the Entropy Model. Secondly, the risk assessment process may be led by the relevant subject matter expert (OSH, environment, quality, operations etc.) but ultimately, the risk owners are those internal stakeholders who hold control of the required resources, systems, and processes, for example, site maintenance manager or site training manager. The major difference from traditional risk assessment processes is that the interdependencies of accountabilities for systems of production risks are much clearer. Finally, risk is presented as having multiple sources. There are eight which are the residual and entropic risks of the four system of production that combine when an activity is carried out. This is more granular than the singular concept that is now used.

DOI: 10.1201/9781032701561-8

CURRENT RISK ASSESSMENT TOOLS

The most used tool for OSH risk evaluation is the Safety Risk Assessment Matrix. This can be a simple five by five table of likelihood versus consequence. Examples can be found online such as *Sample Safety Risk Assessment Matrices for Rail Transit Agencies (dot.gov)* and Risk Assessment in the UK Health and Safety System: Theory and Practice – PMC (nih.gov). These are used to rank risks relative to each other which is commonly referred to as a semi-quantitative method.

The limitations of these risk matrices are well documented. Cox (2008) identified the following deficiencies:

(a) Poor Resolution. Typical risk matrices can correctly and unambiguously compare only a small fraction (e.g., less than 10%) of randomly selected pairs of hazards. They can assign identical ratings to quantitatively very different risks ("range compression").

(b) Errors. Risk matrices can mistakenly assign higher qualitative ratings to quantitatively smaller risks. For risks with negatively correlated frequencies and severities, they can be "worse than useless," leading to worse-than-random decisions.

(c) Suboptimal Resource Allocation. Effective allocation of resources to risk-reducing countermeasures cannot be based on the categories provided by risk matrices.

(d) Ambiguous Inputs and Outputs. Categorizations of severity cannot be made objectively for uncertain consequences. Inputs to risk matrices (e.g., frequency and severity categorizations) and resulting outputs (i.e., risk ratings) require subjective interpretation, and different users may obtain opposite ratings of the same quantitative risks. These limitations suggest that risk matrices should be used with caution, and only with careful explanations of embedded judgments.

(Cox, 2008)

Decision-makers may not be aware of Risk Matrix design features that increase uncertainty. This indicates that the matrix is inherently unreliability as a decision support technique (Peace, 2017).

Matrices, however, continue to be widely used in industry and provide the basis for risk ranking in many enterprise, discipline-specific and operational risk registers. This deficiency was commonly accepted when *PSM1* was written in 2003. As a result, a different approach was developed that considers the breakdown of risks that are then combined into a total risk profile. Whilst the potential combinations of process, technology, physical environment and human resources may be innumerable, the key aim is to get a better understanding of areas of vulnerability or fragility within an organisation. For instance, at a food processing facility, this may primarily be due to aging infrastructure and equipment or reliance on itinerant workers who lack industry knowledge in the agriculture sector.

DYNAMIC COMBINATIONS RISK METHOD

The 'Dynamic Combinations Risk Method' (DCRM) described below provides a mental modelling approach. When a risk professional enters a workplace for the first time, they can use this to identify critical issues, rather than simply looking for hazards. Many hazards in the workplace tend to come and go. For instance, a delivery of supplies is placed in front of a doorway due to lack of space within the workplace. As the supplies are unpacked and put away, the 'hazard' (a blocked doorway) disappears. Whether this is a 'hazard', or an inconvenience depends on whether the doorway could be a critical exit in an emergency. The real risk may be that the facility footprint is too small for the work that needs to be undertaken; however, this is seldom the focus when it's easier to note a blocked exit on an inspection checklist.

Operationally, the Risk Matrix which is often included in tools such as Job Safety Analysis (JSA) is rudimentary at best. Many workers are expected to understand the risk using this likelihood and consequence table. From a strategic perspective too, these are limited for system-wide risk reduction. Constraints can be shown if different activities are compared. For instance, Activity 1 involves an experienced truck driver reversing to the edge of a mine waste dump over which there is a substantial drop protected by an earthen barrier. Using a standard matrix (e.g. page 5 of Sample Safety Risk Assessment Matrices for Rail Transit Agencies (dot.gov)), the probability of an accident would be 'remote' (likely to occur no more than once per five years) and severity 'critical' (a death or partial disability), resulting in a 'medium' risk. Activity 2 has an apprentice using a fixed circular saw in a workshop. The probability of a hand injury is 'occasional' (no more than once per year), and severity is 'critical' (hospitalisation) so the risk is also 'medium'. The risk scores for the two activities are the same even though a truck going over the edge of a waste dump is likely to lead to a fatality or permanent disability (such as paraplegia), whilst the second activity may result in loss of fingers depending on whether the saw has an automatic cut out switch. Examples of similar accidents can be found at: SIR No. 277: Haul truck over open pit wall edge – fatal accident (dmp.wa.gov.au); MSH_SIR_230.pdf (dmp.wa.gov.au) and Teen apprentice loses tips of two fingers – Manufacturers' Monthly (manmonthly. com.au). Incidents from industry help to inform the risk assessment process but don't necessarily make it easier to assign scarce resources for risk mitigation.

It's highly likely that there will be disagreement between participants in a risk workshop about the scores of the two examples above. The matrix is highly subjective so one worker's assessment could differ significantly from another's. Unfortunately, this can lead to debate at the operational level that hinders efficiency. The Risk Matrix doesn't account for the effect of risk modifiers that either increase or decrease the risk such as the level of training received by the worker or the impact of work systems such as shift work. The failure to consider systemic weaknesses that exacerbate risks can lead to resources being allocated to areas which are lower risk. Assumptions must be made about the condition of systems of production. For example, if the truck driver is highly competent (low residual risk), this would reduce the risk. Experience and safety consciousness allow the individual to manage the risks effectively under controlled conditions. If, however, the driver works overtime for consecutive days, the ability to operate safely may be impaired by degradation and the HR entropic risk will rise.

The limitations of the matrix call for a more effective risk evaluation method which allows managers to manipulate the quality of each system of production to pinpoint weaknesses and in turn, increase certainty to improve resource allocation. The Entropy Model indicates that the level of risk in an organisation is dynamic because of the tendency of systems to degrade. Residual risks are assumed to be fixed in the short term due to resourcing constraints and technological limits to HSE in design. Entropic risk, however, can escalate because of changing states or losses of control, whether small, accumulative, or significant.

The DCRM uses a series of scorers for each of the eight sources of risk to generate a more granular risk ranking system. These scorers, like the matrix, work on relative risk rather than absolute risk. Using this method, the first step in risk management is to identify workplace risks for each system of production to better understand the risk profile of an operation. Compiled results are presented in Table 7.1. Different activities are generated by identifying the process, the human resources who undertake the process, the technology used and the physical environment in which it's carried out. The questions to ask are:

- What process is undertaken?
- Who does it?
- What technologies are used?
- Where is the process undertaken?

An activity therefore is the combination of systems of production so that:

Activity = Process + Human Resources + Technology + Physical Environment

Table 7.1 shows three examples which reading across each line, are self-explanatory. The risks within each system of production for these activities can then be identified and semi-quantified using the Scorers explained later in this chapter. To do this, residual and entropic risks are dealt with separately because the two categories of risks require different strategies to manage them, as explained earlier using the Four-Step Risk Management Strategy from the Entropy Model. The identification and semi-quantification of residual and entropic risks leads to Residual Risk Management Strategies (RRMS) and Entropic

TABLE 7.1

Developing Activities Using the Systems of Production

Processes	Human Resources	Technology	Physical Environment
Changing a tyre on a vehicle	Qualified mechanic	Tyre Wheel brace Jack Air compressor	Light vehicle workshop
Building maintenance using a nail-gun to undertake repairs	Experienced maintenance personnel	Nail-gun Nails	Various onsite buildings
Operating a forklift	Qualified forklift operator	Battery powered electric forklift	Warehouse and surrounding yard

Risk Prevention Strategies (ERPS) respectively. There are, therefore, two groupings of risk control interventions using the proposed DCRM method.

RESIDUAL RISK MANAGEMENT STRATEGIES

The development of RRMS begins by quantifying the residual risks within each of the systems of production for a given business activity. Ultimately, the method proposed involves identifying areas of higher risk against those of lower risk for various inputs. For example, an electric saw has a higher residual risk than a hand saw because of the energies involved. An underground mining environment has a higher inherent danger than an office building because it's potentially unstable, confined, has limited illumination and includes other risk factors. Using this method, each system of production is considered in turn to semi-quantify the level of residual risk before combining them again into an activity.

In *PSM1*, the residual and entropic risk scorers included 'Potential Consequences' ranging from catastrophic losses, to severe, serious, moderate and low. In this second edition these have been removed from all scorers except for process because consequences don't directly correlate to the risk level of a single system of production (technology, physical environment, human resources) but to combinations of risks. It's at the interface with processes that these other systems of production contribute to the total risk level of an activity. The scorers in the figures below are therefore greatly simplified from those in *PSM1*. The ranges from 10 (being extreme) to 1 (low) are arbitrary. These are relative so could have been 100 to 1 or 5 to 1. The key is consistency in applying the concepts and focusing the risk workshop audience on understanding how risk sources can interact to create a dangerous and/or inefficient situation.

PROCESS

Every process involves the transfer of energy within the system. Electrical energy, for instance, is used to generate mechanical energy that in turn can be used to project a nail into a piece of wood or crush rocks in a processing plant. The level of energy inherent in a process has the potential to cause negative consequences if released in an unintended manner and lead to injury or damage. This concept was originally proposed in the Energy Transfer Theory of Accident Causation. A worker is injured, or equipment damaged through a change of energy and for every change there is a source, a path and a receiver (Stellman, 1998). The level of energy determines the potential severity of consequences, for example, a spanner falling from a height of ten metres will have a higher level of energy than one that falls from two metres. Process risks are in many ways directly related to the principles of physics and chemistry.

The severity of residual risk therefore depends on the level of energy involved in a process. There are different types of energy that may be present or generated, including kinetic, potential, mechanical, thermal, electrical, radiation and sonar. Each of these has characteristics that can result in losses. For instance, moving machinery parts can cause crush or impact related harm or damage.

To assess the residual risk of the process, the following steps are proposed:

1. Assume that other systems of production are ideal (technology, physical environment, human resources)

2. Define the process e.g. riding a motorcycle at 100 kilometres per hour
3. Assume that the process occurs in static time (a snapshot rather than a movie)
4. Identify the energy involved e.g. kinetic due to riding at 100 kilometres per hour
5. Evaluate the potential consequence of this energy e.g. fatality
6. Allocate a score from 1 to 10 using the process residual risk scorer.

As shown in Figure 7.1, by using energy as the primary risk parameter, the score for process residual risk is linear and not that different from the rationale of the Risk Matrix. It's when other systems of production are introduced that the differences become evident. The basic premise of the process residual risk scorer is that the higher the energy, the higher the risk of the activity in terms of how the work is done. (*PSM1* provided a more detailed explanation of energy types with examples.)

An important rule when using this proposed method is the 'Principle of Separation of Systems of Production'. This is done to give the risk assessor a clearer understanding of inputs. When looking at each system separately, the others are assumed to be ideal – free of residual and entropic risks. For instance, the activity of travelling at 100 kilometres per hour is considered in an ideal physical environment, by an 'ideal' person on a 'perfect' technology. The level of energy, therefore, depends on the speed at which the person or object is going, which in turn, determines the level of residual risk relative to other processes. Using the method of isolating processes from other system of production allows different processes to be compared. In later stages of analysis, other factors are considered, for instance whether the process is undertaken in a car or on a motorbike. The residual risk associated with a particular technology can be assessed, before considering the residual risks of the rider and the physical environment. Figure 7.1 shows the process residual risk scorer in the top panel. The primary risk parameter is the energy level. The potential consequences are listed in the next column from catastrophic to minor loss and these correlate to the residual risk score from 10 to 1.

A key concept in this approach to assessing relative residual risk is that this risk affects the probability of losses independently of time because, according to the Entropy Model, residual risk can't be reduced in the short-term. The process is considered in static time to eliminate the concept of 'exposure'. This is important because residual risk doesn't usually change dramatically in organisational systems unless the enterprise has missed something like unforeseen circumstances or catalysts that act as a trigger. These may be a result of unknown unknowns or incomplete information. It may also be because the circumstances have an inherently high level of residual risk such as when a worker enters an underground mine. The person could be harmed at any time and that's not necessarily a matter of exposure duration. Walking into an unventilated area is forbidden (a 'golden rule') for this very reason. The harm may be almost instant which is why the Entropy Model states that the strategy, both organisational and behavioural, is to remain chronically wary of residual risks (shown in Figure 1.6 as point 3 of the Four-Step Risk Management Strategy).

In the risk scorers, the potential consequences are losses. These could be safety-related, damage to technologies or the physical environment, disruption to production, and/or quality deficiencies. The scale runs from one to ten being low to extreme energy respectively. As stated earlier, the actual numbers are arbitrary and could be one to five, if a consistent approach is taken across all scorers. In the process residual risk scorer, the numbers correlate with the potential consequences of this energy, which range from first aid treatment and/or some damage to numerous

PROCESSES PRIMARY RISK PARAMETER: Energy Level	POTENTIAL CONSEQUENCES	Residual Risk Score
Extreme	**Catastrophic Losses:** Numerous fatalities and/or catastrophic damage and/or Force majeure and/or quality failures with unrecoverable reputation loss	10 9
High	**Severe Losses:** Fatality and/or severe damage and/or Major disruption to production and/or Severe quality failures with significant reputation loss	8 7
Significant	**Serious Losses:** Serious injury/chronic Ill-health and/or serious damage and/or Costly disruption to production and/or Serious quality failures with reputation damage	6 5
Moderate	**Moderate Losses:** Casualty/Ill-health and/or cost-impact damage and/or Disruption to production and/or quality failures with recoverable damage	4 3
Low	**Minor Losses:** First Aid Treatment and/or temporary damage and/or Temporary disruption to production and/or minor failures of quality	2 1

TECHNOLOGY PRIMARY RISK PARAMETER: Energy/ Toxicity Level	RISK MODIFIER HSE in Design	Residual Risk Score
Extreme		10 9 8
High	Adjust downwards for HSE in Design that reduces potential consequences.	7 6
Significant		5 4
Moderate		3 2
Low		1

FIGURE 7.1 The residual risk scorers of each system of production

PHYSICAL ENVIRONMENT PRIMARY RISK PARAMETER: Energy/ Threat Level	RISK MODIFIER Exacerbating Hazards or Threats	Residual Risk Score
Extreme		10
		9
		8
High		7
	Adjust upwards for the presence of exacerbating hazards or threats which increase potential consequences.	6
Significant		5
		4
Moderate		3
		2
Low		1

HUMAN RESOURCES PRIMARY RISK PARAMETER: Average Age by Job Classification	RISK MODIFIERS			Residual Risk Score
	Person / Job Fit	Job-Related Competencies	Risk Management Competencies	
Below 15 years			Adjust upwards if risk management skills training hasn't been undertaken by workers in this or previous workplaces.	10
				9
15 to 19 years	Adjust upwards if new incumbents aren't assessed for job /environment fit.			8
		Adjust upwards according to percentage of workers with less than one years' experience in the industry.		7
20 to 24 years			Adjust downwards if risk management skills training has been undertaken and workers participate in risk assessments about the safety, efficiency and quality of their work.	6
	Adjust downwards if new incumbents are assessed for job / environment fit.			5
25 to 34 years				4
				3
35 years and above				2
				1

FIGURE 7.1   (Continued)

fatalities and/or catastrophic damage. The scale doesn't begin at zero because it's impossible to eliminate residual risk.

TECHNOLOGY

The residual risks associated with technology, which includes plant, equipment, tools, and chemicals, can be determined in a similar manner. Chemicals can be evaluated for the level of residual risk based on toxicity and/or reactivity. The higher the level of toxicity, the greater the residual risk. Plant, equipment and tools can be assessed using the level of energy involved in the same way as processes were analysed, with energy being the primary risk parameter. Technologies, however, as discussed in Chapter 4, can have HSE in design characteristics that reduce the level of residual risk. Logically, these positive risk modifiers should be taken into consideration. A definition can be given to these variables such that:

A risk modifier is a characteristic of a system of production that either increases or decreases the level of residual risk. Negative risk modifiers increase the level of residual risk whereas positive risk modifiers lower this level.

Travelling at 100 kilometres per hour on a motorbike, for example, has a higher residual risk than travelling at the same speed in a car because the structure protects the driver. From a process risk perspective, the residual risk is the same, but the type of technology affects the total risk profile of the activity.

A simple process flow can be used to step through the technology residual risk scoring activity. The first question is what is it – plant/equipment/tool or chemical? If the latter, the residual risk relates to the toxicity or reactivity. For plant and equipment, the next consideration is kinetic energy, specifically whether it's fixed or mobile. If it's mobile, the size as well as the energies involved will be key considerations. If it's fixed, then the scale, types and levels of energy need to be factored in. For both fixed and mobile plant/equipment, HSE in design that either more effectively contains the energy or allows for rapid shutdown, fire suppression or other mitigations to control any potential loss, will reduce the residual risk. For instance, drill rigs that have emergency stops in numerous locations allow anyone working on or near the rig to shut down operations. Another example is onboard automatic fire suppression systems in large earthmoving mobile machines. Features such as these reduce the potential damage or injury from the release of inherent energy sources. Automatic systems that rely on engineered detection rather than human response have a greater impact on residual risk reduction.

The technology residual risk scorer is shown in Figure 7.1 in the second panel. The primary risk parameter is the energy/toxicity level. A line can be extended across from the energy category somewhere from extreme to low, but a downward adjustment should be made for HSE in Design that reduces the potential consequences of injury, damage or loss. This approach, as opposed to the use of the matrix, gives the risk review team, a better understanding of the importance of HSE in Design.

The aim of using risk modifiers is to instil a deep awareness of effective risk management strategies such as HSE in design.

Physical Environment

As explained in Chapter 5, the PE includes natural locational and infrastructure characteristics and can comprise a complex set of risks, some of which may not be completely understood. It's impossible to have a comprehensive dataset especially for natural environment risks. Man-made infrastructure and site designs can have shortcomings causing a poor fit at the interface between these and other systems of production.

The PE residual risk scorer is shown in the third panel of Figure 7.1. Continuing the principles of the process and technology scorers, the primary risk parameter is the energy level. For this system of production, however, the energy may not be quantifiable. For instance, a 100,000-litre tank of unleaded petrol has a known set of chemical characteristics that define the energy being stored. A cyclone/hurricane also has energy, but the intensity can't be accurately forecast. For that reason, the natural PE may better be described in terms of potential threat. Historic data may be useful but can't be relied on. The path of a tornado for instance, may flatten one property and leave neighbouring properties relatively undamaged.

PE risks or threats can work in combination which is worth considering. An earthquake can be accompanied by a tsunami, so the consequences are increased significantly by the double impact. The explosion in Enschede in 2000 was a significant disaster because of the proximity of residences within the blast range. Insufficient lighting underground in an area with cavities substantially increases the inherent danger. The combination of conditions is particularly evident in road use when heavy precipitation fundamentally causes the residual risk to be more difficult to manage. This demands a response of the driver to better manage the residual risk of the process by reducing speed.

The use of the PE residual risk scorer involves identifying the primary energy/threat first and ascribing a severity level, then adjusting upwards if there are exacerbating hazards or risks. For instance, building a nuclear power plant on a fault line, close to the ocean or in the middle of the desert, has very different PE residual risks. In the scorer, additional threats are negative risk modifiers that increase the overall level of risk with the combination affecting potential consequences.

PE residual risks can lead to losses in three ways. The first is by exposing the operation to undesirable energy that puts it at risk, for example, the kinetic energy of the hurricane or the potential energy of earth as it saturates with water and starts a landslide. The second is that the PE can cause loss of control in other system of production, for example, when slippery ground conditions cause a truck to skid or a worker to slip and fall. The third is that the residual risk can be a direct threat to the health and safety of human beings, such as biohazards or smoke from bushfires.

In the scorer, having taken exacerbating conditions into consideration, the PE is then given a residual risk score. These cumulative risks should have been acted on when determining the locations of the nuclear power plants in Japan that were destroyed by tsunamis. Did the operations team fully understand the residual risks they inherited from the project planning, construction and commissioning phases?

HUMAN RESOURCES

In Chapter 6, the residual risks associated with the HR system of production were identified. These were incomplete knowledge, skills and abilities (KSAs) and physical limitations. Whilst the rates of injuries and illnesses amongst older workers have risen and younger workers relatively declined in the last 20 years, it remains true that younger workers have a higher level of risk tolerance and appetite for risk taking. Historically, this was predominantly a young male trait, but societal trends suggest that young women are taking more risks than in the past.

Age and experience have a substantial impact on HR residual risk and are identified in the HR scorer in Figure 7.1 as the primary risk parameter. The risk can be modified through interventions. The modifiers shown provide guidance and each organisation should consider its process, technological and PE risk profile, to determine what HR modifiers would best apply.

The first of these shown in the scorer is creating a fit between the person and the job/work environment from the outset. Many organisations assess this through pre-employment medicals and fitness testing. The assessment should reflect the types of tasks done on the job, for instance, manual handling – lifting weights, grip tests and aerobic fitness for labour intensive roles. This type of evaluation isn't critical in a low-risk office environment but in high-risk environments that are physically demanding, they're a must. This is to ensure the prospective worker is at less risk of injury or harm due to a poor fit based on individual factors. The military and enforcement agencies have used these types of assessments for decades.

The next modifier shown is job-related competencies. If the workgroup comprises many new recruits, then the work team should be managed with higher levels of supervision and preferably, placed initially within a controlled environment during the learning phase. An upward adjustment is made if there's a high percentage of workers with less than one years' experience in the industry. The military and enforcement agencies have adopted this approach using training academies. In industry, however, the longer-term view on personnel development is seldom taken and the pressure to get a recruit working as soon as possible often has a sense of urgency, especially for frontline workers in production-centred industries. Being practical, the issue is not so much whether the person can start working but how the recruit's high level of residual risk will be managed.

The final modifier shown is risk management competencies. If the work group profile includes well-developed risk management competencies, then a downward adjustment can be made, however, if the workers haven't received this training, for instance, in fatal risks and controls, then the adjustment should be upwards. Thinking more broadly, a person with the capacity to evaluate risks dynamically and to adjust their behaviour does so as a life skill, not as a workplace-specific skill. This capability takes time to develop and tends to be enduring. It's driven by an underlying understanding of how things work and can play out. When the military approach is considered, highly trained specialist forces, as part of their work, are exposed to extreme risks and yet, they're capacity to survive is high because of well-developed risk management competencies.

Using the HR residual risk scorer shown in the bottom section of Figure 7.1 should lead the risk assessment team to consider the role of resilience in reducing HR residual risks. It also points to some key strategies. These are: firstly, create as best a fit between HR and the other systems of production associated with the job as possible; secondly, ensure new recruits are managed effectively to minimise residual risks during their learning curve; and thirdly, provide personnel with risk-management competencies to build organisational capacity and resilience.

The scorers are designed to be a learning tool for managers, supervisors and risk-based professionals to direct their mindset away from the singular concept of risk and the simplicity of the Risk Matrix. They should focus on systems rather than behaviours and take a multidisciplinary approach rather than the narrow focus of their discipline (such as safety).

Loss investigators should also be concerned with systems and not start investigations with human factors related to the person/s directly involved in the loss event. The risk modifiers are provided as strategies for prevention. These encourage understanding of the key parameters that affect risk levels within systems of production, for instance, comparing one workgroup profile against another. When a risk professional enters a workplace, they should get a sense of the 'big picture' risk profile of the workplace across the systems of production rather than simply looking for hazards.

CASE STUDY: RESIDUAL RISK ANALYSIS

A case study is used below to demonstrate how the scorers may be applied. Admittedly, this is with the benefit of hindsight but there are lessons to learn for prevention of future disasters of this kind. The BP Refinery explosion and fire of 2005 was discussed briefly in an earlier chapter. In the BP Report, the Colombian Accident Investigation Board is referred to because of commonalities between the two disasters. The following quote describes NASA's culture and structure prior to the Colombia disaster:

> Many accident investigations make the same mistake in defining causes. They identify the widget that broke or malfunctioned, then locate the person most closely connected with the technical failure: the engineer who miscalculated an analysis, the operator who missed signals or pulled the wrong switches, the supervisor who failed to listen, or the manager who made bad decisions. When causal chains are limited to technical flaws and individual failures, the ensuing responses aimed at preventing a similar event in the future are equally limited: they aim to fix the technical problem and replace or retrain the individual responsible. Such corrections lead to a misguided and potentially disastrous belief that the underlying problem has been solved (CAIB, 2003).

(US Chemical Safety and Hazard Investigation Board, 2007)

In Figure 7.2, it's shown that the processes undertaken at the BP site involved extreme energy sources, meaning that the residual risk was extreme. From a technological perspective, there were deficiencies in HSE in Design. These included a breach identified by OSHA when they issued a serious citation alleging that nine relief valves from vessels in the Ultraformer No 3 didn't discharge to a safe place. This exposed employees to flammable and toxic vapours (p. 114, US CSHIB, 2007). Changes were made according to the company, but OSHA didn't conduct a follow-up inspection to verify this. The report identifies other design deficiency in the Amoco Regulatory Cluster Project done in 1993 and the NDU Flare line sizing in 2002. Neither were completed to safety in design standards to accommodate capacity due to cost saving decisions. For the 2002 Clean Streams Project, some planned engineering studies weren't completed and again due to cost controls, an essential component, the wet/dry system was never installed. Later due to scheduling pressures, this project didn't follow safeguards that were part of BP's Capital Value Process.

The report contains a section (5.7.4) that explains inherently safer approaches such as flare systems that combust flammable hydrocarbons before they are vented to the atmosphere, to prevent them from becoming a serious fire or explosion risk (p. 119, US CSHIB, 2007). From this information, in the technology residual risk scorer,

PROCESSES PRIMARY RISK PARAMETER: Energy Level	POTENTIAL CONSEQUENCES	Residual Risk Score
Extreme	**Catastrophic Losses:** Numerous fatalities and/or catastrophic damage and/or Force majeure and/or quality failures with unrecoverable reputation loss	10
		9
High	**Severe Losses:** Fatality and/or severe damage and/or Major disruption to production and/or Severe quality failures with significant reputation loss	8
		7
Significant	**Serious Losses:** Serious injury/chronic Ill-health and/or serious damage and/or Costly disruption to production and/or Serious quality failures with reputation damage	6
		5
Moderate	**Moderate Losses:** Casualty/Ill-health and/or cost-impact damage and/or Disruption to production and/or quality failures with recoverable damage	4
		3
Low	**Minor Losses:** First Aid Treatment and/or temporary damage and/or Temporary disruption to production and/or minor failures of quality	2
		1

TECHNOLOGY PRIMARY RISK PARAMETER: Energy/ Toxicity Level	RISK MODIFIER HSE in Design	Residual Risk Score
Extreme		10
		9
High	Adjust downwards for HSE in Design	8
		7
Significant	that reduces potential consequences.	6
		5
Moderate		4
		3
Low		2
		1

FIGURE 7.2 Applying the residual risk scorers to the BP Refinery Disaster Case Study

PHYSICAL ENVIRONMENT PRIMARY RISK PARAMETER: Energy/ Threat Level	RISK MODIFIER Exacerbating Hazards or Threats	Residual Risk Score
Extreme		10 9 8
High	Adjust upwards for the presence of exacerbating hazards or threats which increase potential consequences.	7 6 5
Significant		4
Moderate		3 2
Low		1

HUMAN RESOURCES PRIMARY RISK PARAMETER: Average Age by Job Classification	RISK MODIFIERS			Residual Risk Score
	Person / Job Fit	Job-Related Competencies	Risk Management Competencies	
Below 15 years			Adjust upwards if risk management skills training hasn't been undertaken by workers in this or previous workplaces.	10 9
15 to 19 years	Adjust upwards if new incumbents aren't assessed for job /environment fit.			8
20 to 24 years		Adjust upwards according to percentage of workers with less than one years' experience in the industry.	Adjust downwards if risk management skills training has been undertaken and workers participate in risk assessments about the safety, efficiency and quality of their work.	7 6
25 to 34 years	Adjust downwards if new incumbents are assessed for job / environment fit.			5 4
35 years and above				3 2
				1

FIGURE 7.2   (Continued)

there is no downward adjustment for HSE in Design which indicates an unchanged potential for catastrophic loss.

The next scorer is the PE residual risk scorer. The actual residual risks of the site didn't appear from the investigation report to be problematic; however, there were several incidents prior to the disaster that required the surrounding community to take shelter. There were two fatal safety incidents in 2004 and a furnace fire for which a community order to shelter was issued. On 28 July 2005, a major process-related hydrogen fire resulted in a Level 3 community alert. This was followed by a gas oil hydrotreater incident that led to another community order (pp. 237 and 238, US CSHIB, 2007).

There were also issues with the location of various hazardous processing plant on the site which went against HSE in Design requirements. This made the physical work environment higher residual risk than needed to be. For instance, the blowdown drum and stack didn't meet engineering specifications for spacing and location of refinery process unit equipment (pp. 112 and 113, US CSHIB, 2007). Site trailers used as temporary office space and to house tools and other equipment, were located between the NDU and ISOM, primarily for convenience. When the disaster occurred, this resulted in more personnel in the area than should have been. In the report, a site diagram is provided in Figure 15. One of the learnings was that all occupied trailers should be located away from vulnerable areas, including relocation outside the refinery if necessary (p. 129, US CSHIB, 2007). As a result of the lack of HSE in Design of the physical work environment, the PE residual risk scorer in Figure 7.2 identifies this deficiency as an exacerbating hazard or threat. The residual risk for this system of production was certainly a contributor to the catastrophic losses.

Detailed information about the risk profile of the workforce isn't provided in the report, except for a few key individuals. The Day Shift Board Operator had six years' experience, two of the five outside operators had over 15 years in ISOM and other operators had nine years (p. 45, US CSHIB, 2007), but some had little experience (p. 46, US CSHIB, 2007). Prior to the disaster, two process technicians with required capabilities weren't assigned to assist with the start-up. As an overall comment, whilst the necessary competencies were available, these weren't assigned to do the critical work prior to the disaster. It can be implied (based on this limited information) that the HR residual risks were on the lower end, but due to ineffective allocation of the work, the HR entropic risk was high. (This will be discussed further in the next section.)

In Figure 7.2, some assumptions must be made due to information gaps. This begins with the age of key personnel. The experience levels of workers and contractors are assumed for the purposes of this exercise to be close to 35 years of age. A further assumption is that some form of personnel assessment was undertaken to ensure that workers were suitable for the job (perhaps, this would have been driven by industry standards). The Report indicates that training was generally deficient so in the scorer, these competency gaps raise the HR residual risk score pushing the potential consequences for HR gaps upwards. A section of the report is dedicated to inadequate operator training including for abnormal situation management. The risk management training was poor as simulators weren't used to prepare operators for hazardous scenarios. They were ill-equipped in this regard, and this was primarily due to failures of the organisation to manage these HR residual risks.

The scorers in Figure 7.2 highlight the key areas of residual risk associated with this disaster. These are:

• Extreme process related energies with catastrophic potential;
• Extreme technological risks related to energies and lack of HSE in Design;
• Moderate PE risks significantly exacerbated by lack of HSE in Design in the layout and use of the site footprint and also the proximity of the community outside the operational site;

- Moderate HR risks greatly exacerbated by lack of risk management competencies to deal with potential threats and losses of control in other systems of production.

The case study will be continued later in this chapter to identify the entropic risks that became evident through the incident investigation.

TOTAL RESIDUAL RISK SCORES

Leaving the case study aside for the moment, once each of the systems of production has been scored for the level of residual risk, these can be combined into activities. An activity involves workers undertaking a process using technology in a physical workplace. This can be represented as an equation:

Activity = Process + Human Resources + Technology + Physical Environment

A suite of activities can be compared by assigning the scores to get a total residual risk score for each activity. The formula is:

Total Residual Risk Score = Prr × HRrr × Trr × PErr

Multiplication is used rather than addition because major deficiencies, for instance, in HSE in Design, will have a significant impact on the total residual risk. For instance, a highly trained military team equipped with the best technologies, when entering a physically dangerous environment, trekking with little or no risk of enemy encounter, will still be exposed to high risk. An example could include hiking through jungle in a foreign land. The multiplication results in an exponential effect when high residual risk systems of production are combined. For instance, allowing apprentice mechanics to undertake complex, hazardous tasks such as removing large components from a piece of machinery using block and chain lifting equipment.

This proposed method of activity-based risk analysis is more comprehensive than approaches using the Risk Matrix and 'What If Analysis' because the focus is on understanding the work as it's done. With this approach, it may be possible to develop a system-wide residual risk profile for a workplace and in particular, high-risk combinations to be identified.

When the Total Residual Risk Score for an activity is calculated, the result can range from 10,000 (10×10×10×10) to 1 (1×1×1×1). Once all the scores for different activities in a workplace are calculated, these can be graphed to identify peak risks. High-risk activities should be given greatest priority. According to the Four-Step Risk Management Strategy, resources should be allocated to ensure that such risks are effectively managed in the short-term and reduced in the longer term. For instance, the organisation may find that its processes are of such high residual risk and are pervasive to the extent that the best method of risk reduction is to change the way a process is carried out so that workers are removed from danger. Examples of where this has occurred include the use of robots or remote-control technologies to carry out hazardous work. A medium-sized haulage company that uses vehicle-mounted cranes may retrofit remote controlled systems that allow the operator to work from a safe distance and have a better, unobstructed view compared to a manually controlled system installed on the side of the truck.

An enterprise may need to consider significant systemic changes if one or more systems of production has a major residual risk that is difficult to manage, especially if a technological solution is available to remove the worker from the hazardous situation. For example, in operations with an extreme PE residual risk, extensive investment in standard technology, HR capability and work practice improvement won't necessarily compensate for the PE residual risk, unless the interface between workers and that environment is changes. Subsea maintenance on oil rigs is an example. In the North Sea, the residual risk associated with the 'workplace' is extreme. If divers undertaken maintenance, investment in safer equipment, training and procedural controls can't mitigate these risks completely. If, however, the worker is isolated from immediate danger and the task carried out remotely, then the residual risk is reduced substantially. The process residual risk score decreases because the worker is no longer directly at the interface with the physical environment. The technology score drops because of HSE in Design. The HR score declines because the work no longer involves physical demands. Only the PE score remains the same. Changing the way in which a process is carried out can therefore greatly reduce the total residual risk. Liu (2022) wrote a paper on the use of remotely operated vehicles (ROV) in subsea engineering which explains these current and future possibilities.

Large organisations with mature systems tend to carry out detailed risk assessments. Quite often these are mandated for insurance processes and as a result, drive their operational risk controls. Other less resourced businesses benefit from regularly carrying out a self-assessment (as discussed in previous chapters). Whilst risk management is compulsory in jurisdictions with HSE legislation, it's also good business practice. The challenge for most is where to start and often the obstacle is being too close to the operations to see the real issues and opportunities. A good starting point is to have open discussions about the safety, efficiency and quality issues that cause frustration in the workplace. It's from these conversations that managers and supervisors may identify low-cost ways of reducing residual risks. The key with known residual risks is to manage them effectively in the short-term but also to envision safer, more efficient ways of working one year, two years and further into the future of the enterprise.

Many SME owners and managers believe capital investment for safety is expensive. What they should be considering is how efficiency can be improved concurrently to maximise return on that investment.

There are other benefits that are often overlooked. For instance, an SME manager should showcase the business' improvements to current and potential clients to build their business brand. An overhaul of the maintenance workshop following Lean principles with a place for everything in its place can be demonstrated with before and after photos. Clients are often interested to hear about safety, productivity and quality improvements because it provides them with reassurance that they've engaged the services of a competent, responsible contractor. The feedback from the client can be taken back to the workers who implemented the workshop improvements.

Focusing solely on the costs of residual risk reduction is a tunnel-vision mindset that holds businesses back from success.

Managers should showcase these investments and any improvements in safety, production, and quality, to stakeholders such as clients.

They should be thinking longer-term about the business' total value proposition.

Importantly, the people who do the day-to-day work should be included in celebrating the 'Felt Success' of the business, especially when there's positive feedback from senior management, clients, regulators and others.

This discussion has highlighted that every organisation, no matter how large or small, should have a strategy for residual risk reduction. A narrow focus on cost keeps businesses in a cycle of losses such as inefficiencies, safety incidents, quality deficiencies, despondent personnel, and frustrated clients. Organisational leaders with an open-mind to opportunities explore tangible benefits and positive flow-on effects of business improvement. For any professional whose job is to put forward improvements to senior management, it pays to prepare a case that includes 'hard' business benefits and 'soft' leveraging opportunities that build reputation and potential future earnings.

ENTROPIC RISK PREVENTION STRATEGIES

As discussed in previous chapters, operations inherit residual risk from the project planning, construction and commissioning phases of a site or facility. The quality of the inputs they inherit also affects an operation's susceptibility to entropic risk as systems of production degrade with time.

Oswald et al. (2020) wrote about construction as an industry structured for 'unsafety' and explored the cost-safety conundrum in project delivery. It's worth reviewing some of these comments and interchanging 'losses' for 'accidents' whilst considering industry more broadly. They identified three costs of prevention:

- Fixed prevention: costs typically being incurred on plant and equipment before production begins and exist regardless of the accident rate.
- Variable prevention: costs proportional to accident frequency rates and their severity. They are linked to the time taken for accident analysis, cause identification and implementation of corrective measures.
- Unexpected prevention: costs initially unforeseen such as an equipment modification to lower noise levels, or changes in legislative requirements or cultural norms.

(Oswald et al., 2020)

The Entropy Model explains that the focus with entropic risk should be on prevention of losses so it's important that organisations better understand their sources of entropic risk to prioritise protection programmes. With the model, mitigation starts with taking timely corrective action to halt escalation. The next step is to review trends and implement proactive maintenance practices. Entropic risk can be variable so can't be semi-quantitatively assessed in the same way as residual risk which is relatively fixed due to technological and resourcing constraints. A methodology was proposed in *PSM1* with the formula for total entropic risk of an activity being:

Total Entropic Risk Score = $P_{er} \times T_{er} \times PE_{er} \times HR_{er}$

For instance, two companies can be compared for their vulnerability to entropic risk. Company A has state-of-the-art technology and a relatively inexperienced workforce. Company B has old technology and experienced personnel. The former would be susceptible to HR entropic risks and needs to invest in capability development whereas the latter must address the wear and tear of its technology as a priority. As was the case with residual risk analysis, the entropic risks associated with each system of production should be considered independently of each other. This again applies the Principle of Separation of Systems of Production.

PROCESSES

In Chapter 3, it was shown that the sources of entropic risk in processes occur at the interface with other systems of production. Vulnerability to degradation depends initially on the quality of inputs (e.g. plant, equipment and infrastructure components) determined at the project planning phase and whenever modifications or changes occur. The second consideration is the fit between process design (how the work is done) and the physical and human assets of the enterprise.

It's worth noting that preventing entropic risk isn't simply a case of building reliability into systems of production. According to Bourassa et al. (2016) in relation to the technology system of production, there are many studies on optimisation of production and maintenance activities. There are gaps, however, as studies on the effects of reliability are rare. As a result, safety and reliability are considered distinct, not interdependent. The researchers indicate that a system can be reliable and concurrently unsafe, and vice versa. It's agreed, however, that good maintenance increases efficiency and lowers incident rates.

What is important is the extent and thoroughness with which the organisation uses activities, such as inspections, incident investigations and other data gathering processes, to take corrective action and to implement maintenance strategies. Fundamentally, entropic risk is prevented through risk management that becomes tangible in operational practice and behavioural habits. These should focus on keeping systems of production at optimised performance.

An enterprise can assess its resilience against degradation by evaluating the effectiveness of its operational feedback and response systems, and KPIs for programmes that maintain or improve the quality of its systems of production.

Entropic risk is variable and difficult to quantify at any given point in time or over a period using the Risk Matrix. The matrix will simply lead to the conclusion that wear and tear of equipment or fatigue in people or corrosion of infrastructure or deviation from safe work processes can occur. In practice, what tends to happen is equipment failure, worker fatigue, infrastructure collapse and unsafe behaviours, become separate line items in the enterprise's operational risk register. These are then given a score based on the Risk Matrix and controls assigned to reduce the risk level. Unfortunately, any one of these could be classified as a fatal risk. Ironically, a worker who is fatigued and sitting at the office desk is more likely to risk embarrassment than cause a serious incident. The point is that the risk happens at the interface with other systems of production not in isolation so singular classification can be meaningless without context.

There is another consideration when evaluating entropic risk in processes. This is the dynamic nature of work. Khanzode et al. (2011) studied recurring hazards in underground coal mines and in support of the work in *PSM1*, acknowledged that work system components continuously deteriorate. They stated that the frequency or time between occurrences (TBO) were measures of this deterioration. In response, a monitoring system was proposed for hazard recurrence phenomenon. Mathematical distributions were used (Weibull and Poisson) to define control limits. The situation was alarming because hazard occurrences at the time of the study, were going to continue with similar magnitude in the days ahead unless a major systems overhaul was undertaken.

> This system overhaul may be planned in terms of technology-related, process-related, physical environment-related and human-related factors. (Mol 2003)
>
> *(Khanzode et al., 2011)*

The researchers focused on degradation rather than rising entropic risk in relation to these recurrent hazards. Control limits (LCL – lower control limit; UCL – upper control limit) were used to evaluate the extent of hazardous conditions in an area of the underground mine.

> If the observations fall close to the LCL or show a downward trend, it is indicative of system deterioration for the hazard considered. On the other hand, observations close to the UCL or showing upward trend indicate improvements in the system. However, in both cases the work system is not under control and assignable causes can be identified for such behavior. A work system is said to be in control if the control chart shows only random pattern of the observations.
>
> *(Khanzode et al., 2011)*

This reinforces a key point discussed in earlier chapters. There will be variability in entropic risk and what the work of Khanzode et al. suggests is that random patterns/ trends of observations in relation to hazard occurrences can lead to assignable causes. If these causes are identified and removed, this can lead to immediate improvements in safety. As a practical example, if it's found that tradespersons regularly must run electrical leads across pedestrian thoroughfares because the power source is too far from where the work must be done (an assignable cause), this can be rectified by installing power leads overhead. A further improvement would be to install power points above each workstation. Notably, this improvement would reduce the safety risk but also improve efficiency.

The Process Entropic Risk Scorer shown in Figure 7.3 illustrates complexity as the primary risk parameter. When the work is organised in overly involved layers of procedures, this can tip the system into informal management of the work to expedite results (Hollnagel's ETTO – efficiency is chosen over thoroughness). There are numerous human factors that will determine whether complex systems lead to compliance or at least, attempted compliance. (These human factors, due to their own breadth and depth, are left out of this discussion. Leveson (2012) talks about 'Complexity and Safety'; Dekker et al. (2011) explore the complexity of failure and the implications for safety investigations. These are in the chapter references.)

In operations, complexity is a hindrance. It can be a recipe for failure in terms of compliance, delivery of the work and is counter to a culture of productive, safety management. The risk modifiers shown in the process entropic risk scorer, in the first panel of Figure 7.3, are firstly, simplification and standardisation with an upward adjustment to the risk if there's been no process review to achieve these workable

PROCESSES	RISK MODIFIERS		POTENTIAL CONSEQUENCES	Entropic Risk Score
PRIMARY RISK PARAMETER: Complexity	Simplification and Standardisation	Regularity, thoroughness and follow-through of monitoring		
Extreme	Adjust upwards if processes have not been risk assessed, simplified, and standardised.	Adjust upwards if high risk tasks aren't monitored.	**Catastrophic Losses:** Numerous fatalities and/or catastrophic damage and/or Force majeure and/or quality failures with unrecoverable reputation loss	10 / 9
High	Adjust downward if processes have been risk assessed, simplified and standardised, with worker involvement.	Transfer across if operational leaders are involved.	**Severe Losses:** Fatality and/or severe damage and/or Major disruption to production and/or severe quality failures with significant reputation loss	8 / 7
Significant		Adjust downward if supervisors and leading workers are involved.	**Serious Losses:** Serious injury/chronic Ill-health and/or serious damage and/or Costly disruption to production and/or serious quality failures with reputation damage	6 / 5
Moderate			**Moderate Losses:** Casualty/Ill-health and/or cost-impact damage and/or Disruption to production and/or quality failures with recoverable damage	4 / 3
Low			**Minor Losses:** First Aid Treatment and/or temporary damage and/or Temporary disruption to production and/or minor failures of quality	2 / 1

TECHNOLOGY	RISK MODIFIERS		Entropic Risk Score
PRIMARY RISK PARAMETER: Age / Usage Rate	Preventative Maintenance	Regularity, thoroughness and follow-through of monitoring	
Extreme	Adjust upwards if technologies are not maintained regularly.	Adjust upwards if high risk technologies aren't monitored.	10 / 9 / 8
High		Transfer across if operational leaders are involved in monitoring technologies in their workplace.	7 / 6
Significant	Adjust downward if technologies are maintained regularly.		5
Moderate		Adjust downward if supervisors and leading workers are involved in monitoring technologies in their workplace.	4 / 3
Low			2 / 1

FIGURE 7.3   The entropic risk scorers of each system of production

PHYSICAL ENVIRONMENT PRIMARY RISK PARAMETER: Age / Condition	RISK MODIFIERS		Entropic Risk Score
	Preventative Maintenance	Regularity, thoroughness and follow-through of monitoring	
Extreme	Adjust upwards if the physical workplace is not maintained regularly. Adjust downward if the physical workplace is maintained regularly.	Adjust upwards if high risk physical environment factors aren't monitored. Transfer across if operational leaders are involved in monitoring physical environment in their workplace. Adjust downward if supervisors and leading workers are involved in monitoring physical environment factors in their workplace.	10
			9
			8
High			7
			6
Significant			5
			4
Moderate			3
			2
Low			1

HUMAN RESOURCES PRIMARY RISK PARAMETER Average Age by Job Classification	RISK MODIFIERS			Entropic Risk Score
	Job Demands	Workplace-Specific Competencies	Risk Management Competencies	
Below 15 years	Adjust upwards if job is: Physical demanding and/or Involves long hours and/or Involves shift work and/or Is mentally demanding and involves long commute times	Adjust upwards according to percentage of workers with less than one years' experience in the organisation.	Adjust upwards if risk management skills training hasn't been undertaken by workers in this organisation. Adjust downwards if risk management skills training has been undertaken and workers participate in risk assessments about the safety, efficiency, and quality of their work.	10
				9
15 to 19 years				8
				7
				6
20 to 24 years				5
				4
25 to 34 years				3
				2
35 years and above				1

FIGURE 7.3   (Continued)

outcomes. A downward adjustment is made when the process has been risk assessed, simplified, and standardised with the involvement of the workers.

The second risk modifier is the regularity, thoroughness, and follow-up of monitoring. An upward adjustment is made if high risk tasks aren't monitored, a transfer across is done if operational leaders are involved in the monitoring and follow-up programmes, and a downward adjustment made if supervisors and workers participate in the review. These adjustments point to a few learning opportunities for the risk workshop audience. Firstly, the organisation must have a fatal risk critical control programme in place to reduce the potential for process entropic risks. This, for instance, includes assessing the effectiveness of crucial procedural controls such as permit to work systems. Secondly, operational leaders (who are the risk owners) must be involved in monitoring the processes that occur within their work area. Accountability for follow-up corrective actions and maintenance practices resides with them. Thirdly, the downward adjustment is made for operational participation because this provides a sense check to ensure that gaps don't arise in procedures and other controls. Instructions must reflect how the work is done not how the work is imagined to be done (as per Dekker's WAI versus WAD).

The scorer indicates that process degradation is most likely to occur in complex work practices. When workers have a simple task to perform deviations from safe, efficient practices are less likely to occur. Complex and/or unfamiliar tasks have a higher level of risk for two reasons. The first is that the requirements of the job can stretch the competencies of the workers, and second, the risks involved might not be fully appreciated.

The engagement of and input from workers is a critical piece in process entropic risk prevention. This ties back to the Four-Step Risk Management Strategy of the Entropy Model. The work behaviours that are encouraged in a risk-based culture are:

1. Take corrective action to rectify areas of degradation (and ensure that behaviours don't lead to degradation such as poor housekeeping);
2. Make and implement sound suggestions to prevent degradation, especially for recurrent issues;
3. Manage residual risks by exercising chronic wariness;
4. Make suggestions to reduce residual risk in the longer-term so that management are better informed when reaching risk-based decisions that affect operations.

The first two points are directed at entropic risk prevention whilst the last two are centred on residual risk management. Efficiency can be part of risk-based reviews and job planning activities.

OSH legislation states that work must be done safely.

There are no clauses to say that it can't also be efficient.

Safety and efficiency can be achieved concurrently if entropic risks that lead to loss of control, are prevented.

Every risk-based activity is an opportunity to not only plan work that goes well but also build organisational knowledge and capacity through a deep understanding of how entropic risks arise and how these can lead to losses impacting safety, production, and quality.

TECHNOLOGY

In Chapter 4, it was shown that the sources of entropic risk for the technology system of production included wear and tear and the technology/operator interface. When equipment is used for the wrong purpose, when the physical environment puts machinery under mechanical strain, and when personnel operate it roughly, wear and tear is accelerated. The primary variable that affects a technology's exposure to these sources of degradation is its age and usage rate. This is illustrated in the Technology Entropic Risk Scorer in the second panel of Figure 7.3.

New equipment has a lead-time before it begins to show signs of wear and tear. Old technologies reach the point where the costs of maintenance exceed the benefits of continued operation. In the scorer, preventative maintenance is a risk modifier. Lack of maintenance raises the entropic risk score. Regularly scheduled maintenance reduces the score. If this is supported by a planned, thorough monitoring regime with corrective actions followed through in a timely manner, the entropic risk score is reduced further. A different approach may be required for chemicals which also fall into the technology system of production. For instance, containers and the like will perish with time so this should be considered.

In industries such as oil and gas, big data analytics have been an emerging trend (Mohammadpoor & Torabi, 2020). The information is sourced from data recording sensors in exploration, drilling and production operations. The analysis results in numerous benefits, including reduced drilling time and increasing safety, optimised performance of production pumps, and improving petrochemical asset management. The reliability of these processes is continually being enhanced. Interestingly, Pettinger (2014) gathered data from safety inspections and found these can be used to develop safety predictive analytics. Unfortunately, the study tended to focus on behaviours and compliance. Other studies found that big data about hydrogen sulphide levels can be used to forecast hazard events and disruptions to production (Mohammadpoor & Torabi, 2020). The possibilities are expanding in the industry and likely being introduced into other highly mechanised operations.

Less sophisticated industries can focus on the fundamentals of preventative maintenance, inspection regimes and corrective action close-out to minimise the degradation of plant, equipment and tools. It's good practice to consider the lifecycle of plant and equipment with the view to replacement ahead of reliability and safety issues. A key consideration is to make informed decisions about the residual risks and future entropic risks of any new technologies and consider this against both short-term and long-term costs.

PHYSICAL ENVIRONMENT

The PE system of production includes infrastructure and locational conditions. In Chapter 5, it was explained that the PE suffers natural degradation, and this can be accelerated by a poor fit with other systems of production. For instance, conditions deteriorate more rapidly when operators drive on the edge of the road rather than in the centre of the lane. Technology can cause deterioration when poorly maintained plant drips oil onto the road. Processes can further change the environment in undesirable ways such as when an ore truck is overloaded causing damage to haul roads or destabilisation of ore tipping ramps (as well as additional wear and tear on the truck).

The rate of deterioration of locational conditions can be difficult to quantify A significant variable in entropic risk is the suitability of the site for the purpose and the quality of site design and planning. During the life of the operation, any design shortcomings that lead to increased degradation tend to become apparent. On sites, a slope may become unstable, and a hollow may fill with water due to poor drainage. All these issues add to operational work and may increase risks. A further example is the 'concrete cancer' and corrosion of infrastructure in coastal areas. The life stage of the operation provides some indication of potential entropic risk. The PE Risk Scorer, shown in the third panel in Figure 7.3, uses the age of infrastructure or condition of the PE as the primary risk parameter.

The same rationale is applied as was done for technology. The scorer shows two risk modifiers which are again preventative maintenance with regular, thorough inspections. Corrective actions are followed up in a timely manner to either rectify degraded states and/or prevent future degradation. A downward adjustment is made for a proactive preventative programme and an upward adjustment if this isn't in place. Inspection regimes ensure that infrastructure and the work site are safe and reliable thereby mitigating the risk of rising entropic risk and failures.

Examples of bridge collapses due to corrosion can be found on the internet. In salt mines, these risks are ever present, and the inspection regimes need to be rigorous, however, some corrosion might not be visible, for instance, to the internal surfaces of ship loading facilities. The collapse of a conveyor system could result in fatalities and would certainly halt supply to customers, and in many jurisdictions, be reportable to the regulator. Because of the high risks and potential consequences involved, salt mining companies should undertake planned non-destructive testing (NDT) of infrastructure and plant in accordance with engineering recommendations. There are several methods that can be applied to evaluate the degradation of materials due to the environment according to various types of corrosion. These include ultrasonic, radiographic, thermographic, electromagnetic and optical NDT (Singh et al., 2019).

As a broader lesson for industry, including SME owners and managers, it's naive to assume that infrastructure will last forever. Most structural building products deteriorate with time. Both a short-term preventative approach to maximise the life of infrastructure should be implemented along with a longer-term replacement or upgrade strategies. The natural PE also changes. For instance, the second-largest wildfire in California was sparked when power lines contacted a tree (CNN, 2022), highlighting that mother nature continues to shift infrastructure towards degraded conditions. The lesson is that all parts of the operational site should be inspected on a regular basis and timely corrective action taken if conditions deteriorate away from a safe state.

HUMAN RESOURCES

In Chapter 6, the entropic risks associated with the HR system of production were identified. These included the new employee who is more likely to deviate from standard practices and optimal behaviours. Other entropic risks that were discussed included fatigue and ill health. These may be due to individual characteristics or induced by work-related systems such as shift work, long hours and demanding tasks. The probability of losses also increases with excessive job requirements that cause stress or difficulty in maintaining concentration. Earlier accident causation models supported the notion that some HR profiles have a greater tendency towards risk-taking behaviour, and this may lead to intentional safety deviations compared to others. As discussed previously, this needs to be put into context within the system especially when undertaking an incident investigation. The behaviours of the individuals directly involved should be the last area of interest after system deficiencies have been thoroughly assessed.

The HR Entropic Risk Scorer shown in the last panel of Figure 7.3 assumes that older workers are lower risk than younger workers who have less developed competencies and lower levels of risk awareness. Accordingly, the age of personnel within a job classification affects not only their level of residual risk but also their tendency towards entropic risk. The Scorer defines average age by job classification as the primary risk parameter.

In the second column, risk modifiers are taken into consideration. An adjustment is made upwardly for jobs and work practices that place high demands on workers. This includes physically and mentally demanding tasks, long hours, shift work and lengthy commuting times. Some personnel who work in remote areas on a fly-in fly-out basis, for example, must get up very early to catch a plane before starting the first shift of the cycle. Before they start work, there is preparation time, travel time to the airport, flight time and travel to the work site, before duties commence. Similarly, personnel who drive long distances between home and work are at greater risk of fatigue and therefore an increased probability of an accident whilst at work and whilst travelling. Job and travel demands are included as negative risk modifiers. (It should be noted that many employers' insurance policies don't cover the workers' commute so whether travel time is a consideration will depend on the enterprise.)

The second risk modifier shown is workplace-specific competencies. Each organisation and workplace are different. Even if businesses are in the same industry, the variations can be significant which requires new incumbents to go through a learning curve or adaptation period. For the sake of simplicity, the scorer assumes it takes the average worker one year to gain sufficient experience to appreciate and manage workplace-specific risks and to become competent in meeting both compliance requirements and cultural expectations. A work team of mostly new personnel would warrant an upward adjustment of the HR entropic risk level.

The final risk modifier shown is risk management competencies. These should be developed as a life skill that supports the organisation's risk management objectives.

The emphasis of training should be on transferable knowledge and skills that enable a person to adapt and adjust according to the dynamic nature of the risks of work (and life) activities.

This is an important element of building resilience and capability in an organisation. In the scorer, an upward adjustment is made if this type of training and on-the-job learning opportunities aren't provided to workers and a downward adjustment made if this is an embedded programme or way of working within the organisation.

Each organisation should consider its operational risk profile and adapt the residual and entropic risk scorers to best fit including adapting the risk modifiers. What has been provided in Figures 7.1 and 7.3 are examples that whilst not specifically tested empirically do align with findings of various researched studies.

CASE STUDY: ENTROPIC RISK ANALYSIS

Earlier, the BP Refinery explosion and fire of 2005 report was reviewed to identify residual risks within each of the systems of production prior to the catastrophic loss. In the section below, the entropic risks will be identified and used to complete the risk scorers based on available information.

The report alludes to the complexity of the processes that were undertaken at the BP Refinery, and in particular, losses of control escalating when the work deviated from expected operational performance. This was exacerbated by the failure to establish effective safe operating limits (p. 100, US CSHIB, 2007). In the first panel of Figure 7.4, the initial primary risk parameter in the Process Entropic Risk Scorer is marked as high. A key inference throughout the report is that systems weren't working as expected. The regularity of this suggests that limited effort had been made to review, simplify and standardise processes to ensure consistent safety and reliability. An upward adjustment is made accordingly.

There are numerous references to the fact that serious previous incidents weren't acted on highlighting issues with the culture (pp. 18 and 25, US CSHIB, 2007). Where inspections were undertaken and defects or deficiencies found, these weren't rectified. It can be concluded that process monitoring regimes were inadequate. This leads to another upward adjustment in the risk scorer. The result is that the process

PROCESSES	RISK MODIFIERS		POTENTIAL CONSEQUENCES	Entropic Risk Score
PRIMARY RISK PARAMETER: Complexity	Simplification and Standardisation	Regularity, thoroughness and follow-through of monitoring		
Extreme	Adjust upwards if processes have not been risk assessed, simplified, and standardised.	Adjust upwards if high risk tasks aren't monitored.	**Catastrophic Losses:** Numerous fatalities and/or catastrophic damage and/or Force majeure and/or quality failures with unrecoverable reputation loss	10
				9
High		Transfer across if operational leaders are involved.	**Severe Losses:** Fatality and/or severe damage and/or Major disruption to production and/or severe quality failures with significant reputation loss	8
	Adjust downward if processes have been risk assessed, simplified and standardised, with worker involvement.			7
Significant		Adjust downward if supervisors and leading workers are involved.	**Serious Losses:** Serious injury/chronic Ill-health and/or serious damage and/or Costly disruption to production and/or serious quality failures with reputation damage	6
				5
Moderate			**Moderate Losses:** Casualty/Ill-health and/or cost-impact damage and/or Disruption to production and/or quality failures with recoverable damage	4
				3
Low			**Minor Losses:** First Aid Treatment and/or temporary damage and/or Temporary disruption to production and/or minor failures of quality	2
				1

TECHNOLOGY	RISK MODIFIERS		Entropic Risk Score
PRIMARY RISK PARAMETER: Age / Usage Rate	Preventative Maintenance	Regularity, thoroughness and follow-through of monitoring	
Extreme	Adjust upwards if technologies are not maintained regularly.	Adjust upwards if high risk technologies aren't monitored.	10
			9
High			8
	Adjust downward if technologies are maintained regularly.	Transfer across if operational leaders are involved in monitoring technologies in their workplace.	7
Significant			6
			5
Moderate		Adjust downward if supervisors and leading workers are involved in monitoring technologies in their workplace.	4
			3
Low			2
			1

FIGURE 7.4   Applying the entropic risk scorers to the BP Refinery Disaster Case Study

PHYSICAL ENVIRONMENT PRIMARY RISK PARAMETER: Age / Condition	RISK MODIFIERS		Entropic Risk Score
	Preventative Maintenance	Regularity, thoroughness and follow-through of monitoring	
Extreme	Adjust upwards if the physical workplace is not maintained regularly.	Adjust upwards if high risk physical environment factors aren't monitored.	10
			9
High		Transfer across if operational leaders are involved in monitoring physical environment in their workplace.	8
			7
Significant	Adjust downward if the physical workplace is maintained regularly.		6
			5
Moderate		Adjust downward if supervisors and leading workers are involved in monitoring physical environment factors in their workplace.	4
			3
			2
Low			1

HUMAN RESOURCES PRIMARY RISK PARAMETER Average Age by Job Classification	RISK MODIFIERS			Entropic Risk Score
	Job Demands	Workplace-Specific Competencies	Risk Management Competencies	
Below 15 years	Adjust upwards if job is: Physical demanding and/or Involves long hours and/or Involves shift work and/or Is mentally demanding and involves long commute times	Adjust upwards according to percentage of workers with less than one years' experience in the organisation.	Adjust upwards if risk management skills training hasn't been undertaken by workers in this organisation. Adjust downwards if risk management skills training has been undertaken and workers participate in risk assessments about the safety, efficiency, and quality of their work.	10
				9
15 to 19 years				8
				7
20 to 24 years				6
				5
25 to 34 years				4
				3
				2
35 years and above				1

FIGURE 7.4 (Continued)

entropic risk prior to the incident was at catastrophic levels, which is what eventuated on 23 March 2005.

The blowdown drum and stack were constructed in 1953 indicating aging technologies were still in operation. The Technology Entropic Risk Scorer, in the second panel of the figure, therefore, starts near high based on age and usage rate. Various modifications were considered by management, and some implemented through to the 2002 Clean Streams Project but not necessarily to expected standards. In terms of entropic risks, the potential was high due to deficiencies in the mechanical integrity programme with a 'run to failure' approach common practice (p. 25, US CSHIB, 2007). Technological risks were poorly managed due to gaps in process safety information, maintenance procedures and training, plus deficiencies in the database maintenance programme (p. 25, US CSHIB, 2007). Instrumentation failures were prevalent (p. 130, US CSHIB, 2007). Monitoring regimes for technologies were inadequate, for instance, corrective action wasn't taken to repair the malfunctioning key splitter tower instrumentation and equipment. The level transmitter was also not repaired, and the pressure control valve was inoperable (p. 48, US CSHIB,

2007). These examples indicate a defective programme for monitoring of technological degradation and the failure to follow up on corrective actions. A further upward adjustment is made in the scorer with the result being high entropic risk in this system of production.

One of the key findings of the report was that the layout of site put more people at risk. Entropic risk was introduced by placing trailers used for office accommodation near the processing plant. Initially, it was intended that this practice would be temporary, but it became a permanent arrangement (p. 122, US CSHIB, 2007). A management of change (MOC) process was completed for the siting of a double-wide mobile office trailer but none of the MOC team had been training in the guidance document used for the building siting analysis.

The report also indicates that traffic, including parking of vehicles, hadn't been controlled including during the more hazardous periods of startup, shutdown or other abnormal conditions. It's stated that *"drivers were not restricted for parking or leaving vehicles idling close to a process unit outer battery limit, where the vehicle engine could introduce a potential ignition source"* (p. 141, US CSHIB, 2007). The key issues that were common to technological entropic risk were evident with the physical workplace, so the risk scorer shows the same pattern of escalating entropic risk.

Finally, there are several clues related to HR entropic risks prior to the incident. For this exercise, a few assumptions are made due to gaps in the information in the report. The average age of workers is again assumed to be in the early 30s based on the small sample of workers described in the report (i.e. the same as for residual risk). The HR Entropic Risk Scorer shown in the last panel of Figure 7.4 starts at this level.

Operator fatigue was identified in the report as a significant issue for the Day Board Operator who had worked 12-hour shifts for 29 consecutive days and slept 5–6 hours in 24-hour periods. The Night and Day Lead Operators had worked 33 and 37 days straight, respectively. All worked 12-hour shifts (pp. 88 and 89, US CSHIB, 2007). The discussions expanded on the impact of this fatigue on cognitive fixation. Personnel were reducing pressure in the plant without questioning why the pressure spikes were occurring. The real problem was not recognised. A significant upward adjustment is made in the job demands section of the risk scorer.

New employees were present in the workplace, but this didn't appear to be an overriding factor in the incident. A flat transfer is made in the scorer under workplace-specific competencies. Gaps in risk management capability are a consistent theme in the report. These were evident as compliance deficiencies at all levels of the refinery including management who didn't lead by example, lack of response to surveys, studies and audits that identified deep-seated safety problems (p. 26, US CSHIB, 2007). There was inadequate risk review for management of change (p. 26, US CSHIB, 2007) and specific issues that required pre-start up safety review (p. 47, US CSHIB, 2007). Training was deficient in most areas (p. 162, US CSHIB, 2007) and it may be assumed therefore, that transferable risk management competencies were also lacking. In fact, the report states that these risk awareness blind spots were found in business units across the corporation. Business unit managers didn't understand or control major hazards (p. 184, US CSHIB, 2007). A further upward adjustment is made in the scorer with a resulting high HR entropic risk.

The completed scorers in Figure 7.4 tell a consistent narrative of degraded states in all systems of production prior to the refinery explosion and fire. The report findings indicate underlying root causes related to serious leadership and cultural issues that pervaded the operation. The recommendations therefore targeted these as a priority. The case study provides valuable lessons learned for all organisations, not just those in the oil and gas industry. Degraded systems of production lead to losses.

(Dr Andrew Hopkins (2021) undertook a review of this case study from the perspective of high reliability organisations (HROs) and provides a cautionary approach to HRO programmes. This is also worth reading and can be found at: A-Practical-Guide-to-becoming-a-High-Reliability-Organisation-Andrew-Hopkins. pdf (aihs.org.au)).

TOTAL ENTROPIC RISK SCORES

As explained earlier, the formula for the total entropic risk score of an activity is:

Total Entropic Risk Score = $P_{er} \times T_{er} \times PE_{er} \times HR_{er}$

As was the case with residual risk, multiplication is used rather than addition because major deficiencies, for instance, lack of corrective action and maintenance practices will have a significant impact on the total entropic risk. The BP Refinery Case Study provides evidence of numerous areas of degradation across the systems of production prior to the disaster.

Using the scale provided, the Total Entropic Risk Score for an activity will generate a result ranging from 10,000 ($10 \times 10 \times 10 \times 10$) to 1 ($1 \times 1 \times 1 \times 1$). Once all the scores for different activities in a workplace are calculated, these could be graphed to identify peak entropic risks. According to the Four-Step Risk Management Strategy, resources should be allocated to ensure that entropic risks are prevented well ahead of significant losses being incurred. An organisation may be able to look at the results and identify which of its systems presents the greatest risk at interfaces with the other systems of production.

In the aviation sector, the concept of degradation has been researched for some time. This is referred to as 'Degraded Modes of Operation'. It's been a central theme in the analysis of aviation incidents. Findings uncovered situations where teams of co-workers continued to operate safety critical systems even when key components of their technological infrastructure were compromised (Johnson and Shea, 2007). The researchers identified a culture of 'make do' or 'work arounds' to continue operations.

In practice, this phenomenon of workers 'making a plan' isn't uncommon for many reasons such as those discussed in Chapter 2 in reference to the Alignment Fallacy. The onus is therefore, on managers and supervisors to go and see what's happening operationally for themselves and to ask questions that don't infer blame. This curiosity should particularly focus on these degraded modes of operation whereby, systems continue to operate but below performance expectations. Managers should have a sense of urgent concern because continuity of production amidst degraded states (and rising entropic risk) isn't sustainable. At some point, the system's tolerances will be exceeded.

Lessons can be learned by taking an engineering perspective on these degraded systemic states within an operation. Failures can be degraded or incipient according to engineering definitions.

A failure which is gradual or partial; it does not cease all function but compromises that function. It may lower output below a designated point, raise output above a designated point or result in erratic output. A degraded mode might allow only one mode of operation. If left unattended, the degraded mode may result in a catastrophic failure.

(American Institute of Chemical Engineers, 2024)

They also refer to 'Incipient failure':

An imperfection in the state or condition of hardware such that a degraded or catastrophic failure can be expected to result if corrective action is not taken.

(American Institute of Chemical Engineers, 2024)

The incipient failure alludes to not only entropic risk but also the possibility of an 'imperfection'. This may be an unforeseen residual risk. It's therefore important for senior and site managers, and risk-based professionals, to understand the residual and entropic risks of their operations to avoid losses. Unmanaged, these will eventually affect the bottom line of the organisation.

TOTAL RISK PROFILE

In *PSM1*, the concept of a Total Risk Profile was proposed. Some examples were used that manipulated the residual and entropic risk scores of different scenarios to identify priorities for risk reduction. Overall, as was shown in the BP Refinery Case Study, the key areas that should be focused on at the operational site level are prevention of degradation and management of residual risks. This applies regardless of the size of the enterprise or the industry that it operates in. Organisations should concentrate efforts on these fundamentals of risk management to reduce the likelihood of safety incidents, losses to production and deficiencies in quality.

Effective risk management requires the consistent operational discipline to set quality standards for systems of production and to maintain those standards. Operational discipline is evident when tasks are executed in a structured, deliberate way.

What this means is that corrective action must be taken in a timely manner to avoid 'Incipient Failures' in one or more of the systems of production. There's no room for 'making do' or laissez-faire management. In this regard, the style of management becomes secondary if this consistency is applied. For instance, day-to-day visible enforcement on a short-term construction contract may be an appropriate supervisor approach to achieving safety, schedule, and quality performance. The effectiveness of this style would, however, be specific to the context. In the longer term, for instance, at an established facility, a culture that has embedded operational discipline in how people go about their work is more resilient than strategies for short-term obedience.

WORKER EXPOSURE: THE COMPLETE PROFILE

Because of its importance, this total risk profile approach is again presented to clarify the complexity of worker exposure at any given point in time. The following formula draws together the risk sources from each of the systems of production including the worker's specific inputs.

Worker Exposure at a given point in time =

Level of residual risk in the process, physical environment and technologies

+ Levels of entropic risk in process, physical environment and technologies

+ Level of competencies and physical ability of co-workers (i.e. others' residual risks)

+ Level of co-workers' degradation (i.e. others' entropic risks)

+ Level of worker-specific competencies and physical ability (i.e. their own residual risk)

+ Level of worker-specific degradation (i.e. their own level of entropic risk)

(adapted from Van der Stap, 2008).

This also helps to contextualise the current debate about 'no blame' and procedural fairness when it comes to incident investigations that involve organisational losses. There are numerous risk factors (external to the individual) that need to be considered before anyone targets the person involved as having 'failed the system'. The last four inputs point to human factors; however, the quality and condition of these are also determined by systems of control to a large extent.

SUMMARY

In this chapter, the current practical interpretation and application of ALARP was challenged on the basis that as an operation becomes more established, incident investigations and other lessons learned tend to move into low value thoroughness actions that don't necessarily reduce risks. Instead, the risk may rise due to increasing complexity. The quandary is that managers feel compelled to do something as doing nothing is not an option. There's a tendency to add more administrative controls rather than going back to the fundamentals of effective risk management that involve HSE in design of technologies, the physical workplace and processes (how the work is done). There's an over-reliance on actions directed at improving human performance rather than creating a safer, more efficient workplace. Such administrative controls are an easy option due to lower cost compared to higher order engineering controls.

A significant hindrance to operational risk management is that it's constrained by current risk assessment processes based on the Safety Risk Assessment Matrix. These are more akin to Work as Imagined than Work as Done because it's based on

'What If Analysis' rather than what's happening in the workplace. The interfaces between systems of production tend to be overlooked. An alternative mental model was presented which identifies and assesses the residual and entropic risks in the four systems of production. From these, eight risk scorers were developed which factor in the primary risk parameter such as energy level, and risk modifiers which either increase or decrease the risk level. These take the risk review team directly to the underlying strategies for risk reduction. In addition, the approach is systems-driven (inclusive of HR risks) not centred on behaviours. Each organisation should consider what risk modifiers best apply to their operations.

The BP Refinery explosion and fire in Texas (2007) was used as a case study to apply the risk scorers. The residual and entropic risks were readily categorised enabling strategies for systemic improvement to be identified. It's important that organisations take lessons learned across industry sectors such as oil and gas, rather than confining opportunities within an industry. Residual risk reduction strategies and entropic risk prevention strategies provide opportunities for business performance and improvement in any organisation but are particularly important for high-risk industries that have the greatest vulnerability to catastrophic losses.

The Risk Matrix generally results in the production of a Risk Register comprising multiple line items that describe hazards or risks that can occur in operations. It doesn't provide an understanding of how the work is done. These are intended to be 'live' documents but are in practice, maintained by the OSH department with reviews undertaken annually (quarterly at best). Instead, it was proposed that a Total Risk Profile could be compiled of the key activities undertaken in operations to better understand the interfaces between systems of production and the respective systemic entropic and residual risks. What processes are being undertaken? With what technologies? In which work areas? By whom? What are the known residual and entropic risks within these systems of production and how is this affecting the operation's risk profile? How are the residual risks being managed? How are the entropic risks being prevented? Are the operations suffering losses (safety, production, quality)? Does management have control? Have operations become complacent?

The nature of hierarchical structures is that senior managers become further and further away from the reality of day-to-day operations. They can't get bogged down in the detail, but they must create an environment that delivers assurance. It's generally accepted that top-down management constrains operational resilience, and the Alignment Fallacy took this further to suggest that it's delusional to expect head office policy to materialise into operational reality when there are conflicting objectives and areas of non-alignment that are ignored. Operational capacity needs to be developed bottom-up concurrently. In this chapter, it was explained that senior management should ensure that personnel are provided with opportunities to develop risk management life skills by being active participants in risk assessments, risk-based work planning and other experiential activities. (This topic will be discussed in detail in Chapter 9).

Chapters 3 to 6 dissected the residual and entropic risks in each of the systems of production. In Chapter 7, these were combined to better understand the risks at the interfaces within operational activities. The proposed risk assessment method went back to the fundamentals of effective risk management which are to achieve 'as low as reasonably practicable' tempered by a sense check on the value of adding further layers of administrative controls.

A key underlying theme has been the need to revisit the risk profile that operations inherit from the planning, construction, and commissioning phases of a facility's lifecycle. This entails risk leaders having a healthy wariness of unknown residual risks and a clinician's appetite for monitoring operations for changes that could lead to losses. The view should be through the broad lens of systems that are interdependent and not on operational managers or supervisors or workers as if they're a problem to be fixed. Rationally, people are part of the system. They aren't the system.

In the last three chapters of this second edition, the discussions will turn to building risk leadership at all levels of the organisation. Current enterprises intending to survive long-term amidst the dynamics of the Fourth Industrial Revolution will focus on their capacity to manage risks and exploit opportunities. Traditional ways of managing in disciplinary silos create overlaps, waste, conflict and sub-optimal organisational performance.

An aim of *PSM2* is to contribute to better operational management. It's critical that risks and opportunities are better understood but it's the mindset change from discipline-based (production, OSH or quality) to risk-based thinking that will make the difference. For this reason, the rest of this books talks about risk and opportunity leadership and management.

The final chapter will invite risk and opportunity owners to think strategically with the understanding that every action matters. Newton's third law, "For every action, there is an equal and opposite reaction" may apply to forces between interacting objects, but it doesn't apply to human interactions, unless a mindset of conflict is held by individuals, groups or the organisation. Some get entrenched in this malaise which degrades the culture with time. It takes a strong and visionary leader (preferably, leadership team) to 'turn the ship around'. The discussions in the next chapter will set the scene by revisiting the Entropy Model as it applies to Organisational Capacity.

REFERENCES

American Institute of Chemical Engineers, 2024, Center for Chemical Process Safety, Degraded Failure I AIChE. https://www.aiche.org/ccps/resources/glossary/process-safety-glossary/degraded-failure

American Institute of Chemical Engineers, 2024, Center for Chemical Process Safety, Incipient failure I AIChE. https://www.aiche.org/ccps/resources/glossary/process-safety-glossary/incipient-failure

Bourassa, D., Gauthier, F., & Abdulnour, G., 2016, Equipment failures and their contribution to industrial incidents and accidents in the manufacturing industry. *International Journal of Occupational Safety and Ergonomics* 22, 1–23. https://doi.org/10.1080/10803548.2015.1116814

Columbia Accident Investigation Board (CAIB) Report, 2003. Vol. 1, Washington, DC: National Aeronautics and Space Administration.

CNN, 2022, California's second-largest wildfire was sparked when power lines came in contact with a tree. Cal Fire says. CNN.

Cox, L.A., 2008, What's wrong with risk matrices? *Risk Analysis Journal*, Society for Risk Analysis. –(Wiley Online Library).

Dekker, S., Cilliers, P., & Hofmeyr, J-H., 2011, The complexity of failure: implications of complexity theory for safety investigations. *Safety Science*, 49(6), 939–945. ISSN 0925-7535, https://doi.org/10.1016/j.ssci.2011.01.008

Department of Mines and Petroleum Resources Safety, Government of Western Australia, Significant incident Report No. 230, Dump truck roll-over – fatal accident, MSH_SIR_230.pdf (dmp.wa.gov.au).

Department of Mines and Petroleum Resources Safety, Government of Western Australia, Significant incident Report No. 277, Haul truck over open pit wall edge – fatal accident, SIR No. 277: Haul truck over open pit wall edge - fatal accident (dmp.wa.gov.au).

Hopkins, A., 2021, A-Practical-Guide-to-becoming-a-High-Reliability-Organisation-Andrew-Hopkins.pdf (aihs.org.au).

Johnson, C.W., & Shea, C., 2007, Understanding the contribution of degraded modes of operation as a cause of incidents and accidents in air traffic management. *Engineering, Environmental Science.* https://www.dcs.gla.ac.uk/~johnson/papers/degraded_modes/ATM_Degraded_Modes_v1.pdf

Khanzode, V.V., Maiti, J., & Ray, P.K., 2011, A methodology for evaluation and monitoring of recurring hazards in underground coal mining. *Safety Science*, 49, 1172–1179.

Leveson, N.G., 2012, Complexity and safety. In O. Hammami, D. Krob, & J.L. Voiri (Eds.), *Complex Systems Design & Management.* Berlin, Heidelberg: Springer. https://doi.org/10.1007/978-3-642-25203-7_2

Liu, G., 2022, Remotely operated vehicle (ROV) in subsea engineering. In W. Cui, S. Fu, & Z. Hu (Eds.), *Encyclopedia of Ocean Engineering.* Singapore: Springer. https://doi.org/10.1007/978-981-10-6946-8_225

Manufacturers' Monthly, 2011, Teen apprentice loses tips of two fingers. Manufacturers' Monthly (manmonthly.com.au).

Mohammadpoor, M., & Torabi, F., 2020, Big Data analytics in oil and gas industry: an emerging trend. Petroleum, 6(4), 321–328. ISSN 2405-6561, https://doi.org/10.1016/j.petlm.2018.11.001

Mol, T., 2003, *Productive Safety Management – A Strategic, Multidisciplinary Management System for Hazardous Industries that Ties Safety and Production Together.* Oxford: Butterworth-Heinemann.

Oswald, D., Ahiaga-Dagbui, D.D., Sherratt, F., Smith, S.D., 2020, An industry structured for unsafety? An exploration of the cost-safety conundrum in construction project delivery, *Safety Science*, 122. ISSN 0925-7535, https://doi.org/10.1016/j.ssci.2019.104535.

Peace, C., 2017, The risk matrix: uncertain results? Policy and Practice in Health and Safety, 15(2) (tandfonline.com), 131–144.

Pettinger, C.B., 2014, Leading Indicators, Culture and Big Data: Using Your Data to Eliminate Death. Presented at the ASSE Production Development Conference and Exposition, Orlando, FL, June 2014.

Rail Transit Agencies, 2019, Sample safety risk assessment matrices, sample safety risk assessment matrices for rail transit agencies (dot.gov).

Russ, K., 2010, Risk assessment in the UK health and safety system: theory and practice. *Safety and Health at Work*, 1(1), 11–18. https://doi.org/10.5491/SHAW.2010.1.1.11. Epub 2010 Sep 30. PMID: 22953158; PMCID: PMC3430933.

Singh, R., Raj, B., Mudali, K., & Singh, P. (eds.), 2019, *Non-Destructive Evaluation of Corrosion and Corrosion-assisted Cracking.* NJ: The American Ceramic Society.

Stellman, J.M. (ed.), 1998, *Encyclopedia of Occupational Health and Safety.* 4th ed. Geneva: International Labor Office.

US Chemical Safety and Hazard Investigation Board, 2007, Investigation report – refinery explosion and fire, March 23 2005. https://www.csb.gov/file.aspx?DocumentId=5596 and update at BP America (Texas City) Refinery Explosion. CSB.

Van der Stap, T., 2008, Overcome the conflict between safety and production with risk management and behavioural safety principles, *RM/Insight,* 8(2), 16–20, American Society of Safety Professionals.

8 Risk Leadership

THE ENTROPY MODEL AND ORGANISATIONAL FACTOR RISKS

After the ASSP Conference in 2020, Grieve and Van der Stap, had an article published in the *American Society of Safety Professionals Journal* entitled, "Safety and Entropy: A Leadership Issue". The need for this thought leadership piece arose out of the operational experience of the authors having seen well-performing organisations and teams decline because of negatively impacting external and internal influences. The article asserted that there are four organisational factors critical to sustaining capacity and these have residual and entropic risks.

Before exploring this version of the Entropy Model, it's important to understand 'capacity' and why it's important. There are numerous definitions and the one that resonates strongly and applies to all disciplines is

> that emergent combination of attributes that enables a human system to create developmental value.

> *(Morgan, 2006 and van Popering-Verkerk et al., 2022)*

There are five central concepts of capacity:

1. empowerment and identity, properties that allow an organization or system to survive, to grow, diversify and become more complex. To evolve in such a way, systems need power, control and space.... people acting together to take control over their own lives in some fashion
2. collective ability, i.e. that combination of attributes that enables a system to perform, deliver value, establish relationships and to renew itself. Or put another way, the abilities that allow systems... to be able to do something with some sort of intention and with some sort of effectiveness and at some sort of scale over time.
3. an emergent property or an interaction effect. It comes out of the dynamics involving a complex combination of attitudes, resources, strategies and skills, both tangible and intangible... And it usually deals with complex human activities which cannot be addressed from an exclusively technical perspective.
4. a potential state... It is about latent as opposed to kinetic energy. Performance, in contrast, is about execution and implementation or the result of the application/use of capacity. Given this latent quality, capacity is dependent to a large degree on intangibles... This potential state may require the use of different approaches to its development, management, assessment and monitoring.
5. creation of public value.

> *(Morgan, 2006)*

DOI: 10.1201/9781032701561-9

These concepts will be embedded in the discussions throughout Chapters 8–10. Broadly, it's proposed that enterprises that are effective at managing risk and exploiting opportunities have the following characteristics:

- Mature (in terms of delivery) leadership at all hierarchical levels as appropriate to:
  - current situational conditions (efficacy);
  - foreseeable potential conditions (readiness); and
  - new and possibly unforeseen conditions (adaptability);
- Well-developed competencies that lead to sound performance across multiple business functions;
- Applied, right-sized management systems managed through a multidisciplinary approach with operational risks owned by the relevant risk owner/s; and
- Resourcefulness/resilience enabling risk-based problem-solving and effective decision-making.

The model in Figure 8.1 starts on the left-hand side with the four organisational factors that are critical for success. These are leadership, competencies, management systems and resourcefulness/ resilience. In a perfect world these would be in an ideal state with the level of risk being zero. This unrealistic paradigm is characterised by safety, production output and quality at high performance levels.

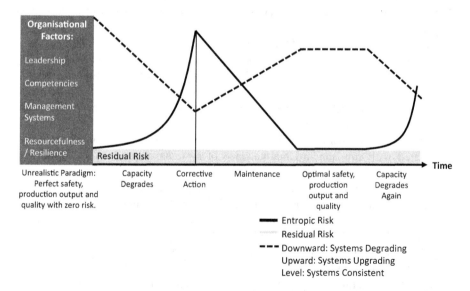

FIGURE 8.1 The fifth section of the Entropy Model but adapted for organisational factors and capacity

In the next section, the residual risk is shown. None of these organisational factors will ever reach perfection as human fallibility has an over-riding influence on all of them. As was the case with the original Systems Approach Entropy Model, this residual risk needs to be managed and again, avoidance of complacency is critical. The figure also shows that these factors can degrade which requires a corrective action response. Any trends alert the organisation to the need for maintenance regimes to ensure that factors are upgraded to achieve optimal safety, production output and quality. With complacency, these factors can degrade again causing a return to rising entropic risk and potential losses.

Throughout the discussions in previous chapters, the importance of managing residual risks and preventing entropic risks of systems of production has been the primary focus. The approach has centred on 'hard' tangible elements of the organisation (including objectively treating HR as an input). These hard inputs can be likened to the enterprise being a wall of bricks, however, without the mortar, there's danger of collapse. The bricks must be held together by factors that build capacity. Where they become deficient, serious issues can arise as identified in the various case studies reviewed earlier. The root causes of many disasters stemmed particularly from poor leadership and unhealthy cultures.

The literature contains extensive research on the importance of leadership, the benefits of strength and depth, and the consequences of poor delivery. There can be numerous causes of degradation overall such as downsizing or turnover resulting in loss of local knowledge, influential relationships, subject matter expertise and critical communication pathways. Personnel can drift to under-performance because of increasing core business pressures beyond their capability to deliver outcomes. External (non-work-related) personal issues can also be a cause or contributor to this degradation. It's not surprising that psychosociological safety, health and wellbeing of workers is now included in OSH legislation.

The Strategic Alignment Channel identified external factors that can significantly impact an organisation. Economic conditions that drive cost cutting including workforce redundancies can result in loss of capacity to do the work to the required standards of efficiency, HSE performance and quality output. This may in turn cause degradation to systems of production such as equipment reliability, workplace conditions and lost investment in HR training and development. The consequences are associated with sub-standard maintenance of systems of production. 'Hard' and 'soft' elements are therefore interdependent.

This loss of human capital may limit the organisation's ability to adapt to change and external pressures. For instance, during COVID-19, many companies faced significant risks because of the inability to maintain personnel levels. Those businesses that had little or no flexibility regarding production timeframes such as agriculture (e.g. harvesting of crops or supply of meat to market) were focused on survival day-to-day, week-to-week etc., not on building resilience. The duration of COVID-19 was a lost opportunity in terms of business improvement for many organisations. It's important for managers to understand where an organisation is in terms of survival versus capacity development and why this is the case.

Capacity can also be eroded when management systems are not maintained to capture the work being done. This degradation can expose the enterprise and its officers to OSH legislative risks. Degraded management systems may be ongoing until resources become available for an overhaul (corrective action phase) and implementation through socialisation of the system (maintenance phase). In the meantime, the gap between work-as-imagined (as described by the redundant management system) and work-as-done (how work is being done presently) may have increased, leaving the enterprise vulnerable to systems 'dementia'.

Degradation of management systems leads to rising entropic risk (characterised by disorder and chaos) culminating in 'Systems Dementia'. The organisation's memory, cognitive skills, and social structures can fade beyond recognition. There's an additional organisational risk if leaders accept this degradation because, in the short-term, the system continues to provide acceptable performance.

Where the management system is no longer the source of 'truth', personnel may do workarounds or take shortcuts to ensure production continues; after all, a management system is only a repository of organisation knowledge within a given period if it's not maintained.

Figure 8.1 can be used to explain the processes for preventing loss of organisational capacity. These follow the Four-Step Risk Management Strategy:

1. Take corrective action to address areas of degradation (and rising entropic risk);
2. Use data to identify trends and implement maintenance practices to prevent degradation;
3. Manage residual risks that are due to human fallibility embedded in organisational systems;
4. Reduce residual risk in the longer-term through capital investment including technological solutions.

It's worth revisiting the five central concepts of capacity to see how these fit with these organisational factors. The following are proposed as triggers for strategy development:

1. Empowerment and identity can only be enabled through effective leadership so invest in risk leadership skills;
2. Collective ability is the equivalent of the total competency profile and the quality of those competencies within an organisation so invest in bottom-up, participative strategies;
3. The emergent property is synonymous with resourcefulness/resilience and essential for adaptability to change so identify and develop internal expertise to mitigate risks;
4. Tangible 'hard' system factors and intangible 'soft' factors need to be developed, managed, assessed, and monitored which require a management system that is sufficiently robust but flexible to enable both to be aligned; and

5. The creation of public value is supported by preventing losses and considering stakeholder needs so know who these stakeholders are and their expectations.

The key strategy development areas are therefore: leadership development programmes; competency profiling and development; robust management of change programmes; mature management systems with hard and soft factors; and stakeholder needs analysis.

Not all organisations will have the resources to deliver these strategies, however, if senior management begin with awareness of the risks, that's a good start. Likewise, some businesses struggle to achieve a low baseline level of residual risk as this risk may be determined by labour market variables such as availability of needed managerial, supervisory, skilled and semi-skilled personnel. A tight market with high demand for skilled workers will tend to constrain businesses that can't meet remuneration expectations. As a result, such enterprises tend to have higher HR residual risk because of their inability to attract and retain high value human capital.

The level of residual risk is fixed in the short-term due to time and resource limitations to develop, implement and embed capabilities. For instance, the agriculture sector in some countries relies on tourists with working visas to support harvesting. Many of these personnel don't have experience and should be supervised closely to manage the risk. There's also no incentive to invest in risk management skills of this labour force because of the short-term, itinerant nature of the work.

Companies that struggle to compete for leadership skills and workforce competencies need to better understand their capacity constraints and develop strategies for short-term management and longer-term improvement. Similar risk profiling can occur, however, in larger companies depending on the phase of the facility's lifecycle. As discussed earlier, construction often requires a sizable workforce for limited time. There is little opportunity or financial incentive to instil risk management competencies, influential leadership and values related to safe, efficient, quality work with short-term workers. Organisational capacity is often limited to essential trade and labour skills at the work-front level. The residual risk can therefore be high compared to steady state, established operations. Low investment in capacity building can be readily justified. Management of the gaps associated with this type of work often comes down to highly visible, on-the-ground supervision driven by mandatory rules and compliance.

Command and control management, whilst not in vogue, isn't wrong.

It depends on the risk profile of the work and the capacity of the workforce to manage the risks. More conscious effort is needed to understand and manage capacity constraints and HR risks.

The model in Figure 8.1 shows that the four organisational factors can degrade with time. This may occur singularly, in combination or collectively. A singular example could involve the loss of a positive and influential leader from an operational site. A collective consequence may ensue if there's subsequent attrition of that leader's followers. Such a situation may particularly weaken the culture if new incumbents feel pressure to prove their worth by pushing short-term production targets to the detriment of the quality of systems of production. The gap may be the lag time to assimilate these new recruits from a psychological contract perspective.

An ongoing decline may result from continued loss of these internal competencies, with recovery requiring a long time. Integrating new personnel into an organisation doesn't happen in days or weeks. Time isn't just required in terms of job skills but also socialisation into how and why the organisation operates as it does. Why does the enterprise have a 'risk- and opportunity-based culture' as opposed to 'Safety First'? Why is everyone asked to contribute by taking timely corrective action and suggestions for maintenance practices? Capacity is built with time – with investment in people. When leaders leave, capacity departs with them. Those with a strong, positive influence can create a significant gap and yet, many organisations fail to do more than just an exit interview with the departee. A risk-based perspective is seldom taken because it falls into the 'undiscussables' where organisational silence is preferred over truth.

Serious degradation of these non-physical factors can lead to significant incidents or losses by affecting the quality of risk-based, decision-making, as was seen in the Challenger and Columbia disasters. In industry, management's tolerance of risk may shift towards higher acceptance as spending cuts impact maintenance budgets and flow onto reduction in plant and equipment reliability, infrastructure/ workplace conditions, the process controls ordinarily in place, and the ability of workers to deliver to required standards. Fewer workers may be expected to generate the same production resulting in longer hours and higher productivity expectations per person. As the entropic risk rises, the likelihood of losses increases. If safety incidents or other losses become evident, management should be seriously concerned. This should be a trigger to ensure operations are being controlled, otherwise, there's a risk not only of internal wastes and inefficiencies but also of events that cause reputational damage and/or loss of critical external relationships (clients, customers, community, government etc.) to the extent that the organisation may no longer be considered a reliable supplier or contract partner.

These matters related to management of capacity tend to be overlooked by organisational leaders and are outside the scope of most current business risk assessment practices. Operationally, the only influencing factors may be recent changes in OSH legislation in some jurisdictions resulting in a growing focus on psychosocial hazards and risk management. This will contribute to greater concern for individuals (whether in management or on the shop floor) in relation to mental health and wellbeing. However, whether this translates into a more concerted effort from a broader risk perspective is uncertain. For instance, the decline in leadership capability of a senior manager due to professional and/or personal factors should trigger early intervention as both an OSH duty to that person and ahead of any subsequent psychosocial risk to that person's peers and subordinates.

An important message to be drawn from Figure 8.1 is that like the System Approach Entropy Model, the enterprise, having achieved 'optimal safety, production output and quality' may suffer degradation of capacity again and therefore, the cycle might be repeated. Other issues or minor losses may be evident including HR problems, poor governance, attrition of key personnel or other events that may predispose the organisation to decline. Behind these concerns is the need for better management and leadership.

Some of the major industrial disasters that have occurred over the last 50 years indicate that the companies involved hadn't recognised their exposure to organisational factor entropic risk.

For instance, at BP's Texas City Refinery which exploded in 2005, the United States Chemical Safety and Hazard Investigation Board explained that production pressures and cost-cutting programmes were a root cause of the accident. Management had relied on injury data as evidence of safety performance and weren't aware of the poor safety culture at the plant (Isiadinso, 2015).

In 2014, DuPont had a chemical release that killed four workers. The chairperson of the Investigation Board indicated, *"What we are seeing here in this incident in LaPorte is definitely a problem of safety culture in the corporation of DuPont"* (Hosier, 2015). In both cases, the Entropy Model may have prompted an investigation into degraded states within systems of production and importantly, organisational factors. What can happen in the enterprise is that 'safety' can be too easily pigeon-holed into a set of key performance indicators (KPIs) and a silo rather than being fully integrated into management systems, practices and supported by a robust, risk- and opportunity-based organisational culture.

In these and other cases, senior managers weren't aware of the degraded state of the business' systems or organisational capacity until after the catastrophe occurred. Well ahead of such disasters, the enterprise may have been suffering losses in production/productivity/schedule, quality deficiencies and/or safety incidents (perhaps unreported) from degradation. Real risks at the operational level need to be recognised and articulated to create a sense of urgency about vulnerability to failure. This requires that leaders make a step-change from safety- to risk-based thinking within the company and more broadly at the interface with its key stakeholders, such as contractors and suppliers. It should be remembered that 'safety-thinking' tends to be marginalised in managers' and supervisors' decision-making processes when core business objectives – production targets, construction schedules, customer satisfaction, patience care, revenue targets etc. – are slipping.

'Safety thinking' is not sustainable when an organisation comes under pressure. A shift is needed to 'risk- and opportunity-based' thinking and development of capacity.

SITUATIONAL LEADERSHIP FOR OPERATIONS

In the earlier discussion, it was stated that strict operational control management practices may be suitable for short-timeframe projects such as agriculture and construction. This clearly calls for a different leadership style to that required for operations which benefit from longer-term, capability development strategies. Construction projects demand problem-solving that reverts to technical expertise, planning and supervision of interfaces between contractors, visible enforcement especially of fatal risk critical controls and a culture characterised by directive regulation in a dynamic environment. Command and control shouldn't be confused with 'dictatorship' or other negative connotations.

PSM1 included an adaptation from the Tannenbaum-Schmidt Continuum (Tannenbaum and Schmidt, 1973) with the addition of the risk level from high to low and the respective timeframe for action, ranging from urgent to not urgent. (This is described here without the inclusion of the figure.) High risk, urgent work demands an assertive leadership style where the manager makes the decision and announces it. This is appropriate in emergency situations and hazardous short-term projects. The use of supervisory authority is critical and should be supported by the hierarchy. A 'hard' style doesn't mean that there's no care or concern for workers. Values that underpin preventative safety measures may be communicated differently, perhaps without 'soft' skills. The reality of schedule pressure often constrains collaboration with workers so essential risk management communication and participation tends to be applied through command and control. The right of workers/anyone to stop unsafe work must be mandated and acted upon to ensure timely corrective action and prevent losses, but also to convey the criticality of safety as a project deliverable.

At the other end of the continuum is the low risk, non-urgent working environment. The leadership style becomes increasingly collaborative with opportunities for consultation. For the organisation to build capacity and resilience in the longer term, this should occur through risk management practices, work planning and other activities that apply a decentralised approach to risk ownership. The appropriate leadership style is situational. To achieve safe production, there needs to be inherent resilience in an organisation's capability adapting leadership delivery to changing circumstances.

In the middle ground between assertive and collaborative leadership, the manager or supervisor has numerous strategies when engaging with people in decision-making. According to the continuum, they may elect to 'sell' a decision, present a tentative decision subject to change, or define limits and ask the group to make the decision. The approach taken depends on numerous dynamics, but the primary considerations should be the level of risk and urgency with which decisions need to be made.

Overall, the most successful leadership style is based on situational factors. In high-reliability organisations (HROs), one of the five principles of effectiveness is 'Deference to Expertise'. During high paced operations when there are unexpected problems, decision-making migrates from management levels to the lowest levels where a team is formed with the specific knowledge, skills, and expertise to deal with the event. Hierarchical rank is subordinated to expertise and experience. The inference is that HROs are more concerned about managing risks and implementing well-informed actions than the politics of managers always being the decision-makers and being perceived as leaders. Credible leaders understand their limitations and confidently delegate to those with well-developed competencies in specialised fields.

The concept of organisational capacity lends itself to a more objective approach to leadership. Executive and senior management teams have residual risks (including limits to their knowledge and skills). Collectively, these residual risks can include organisational groupthink. This is the tendency of teams to select decisions based on consensus. Symptoms include beliefs of invulnerability; relying on rationale characterised by ignoring warnings; faith in the morality of the group; stereotyping of those who are outside or challenge the group; self-censorship such that individuals keep silent; pressure on anyone who expresses doubts; the illusion of unanimity; and mind guards who protect the leader and other members from adverse information (Janis, 1971).

Negative group traits that obstruct sound risk- and opportunity-based decision-making are a residual risk. Within this phenomenon, one of the symptoms of groupthink is stereotyping. This is commonplace as a human behaviour in society and the workplace. As a residual risk, this needs to be consciously managed. A typical operational example is the person in the group who engages in devil's advocacy dialogue, often pointing out the flaws of a decision or how work is done. It's a mistake to stop listening to these personality types because they break with consensus and may identify genuine flaws others don't see or fear speaking up about. The dangers of groupthink are evident in many disasters. Blind optimism and miscommunication within the organisation led to NASA's desire to complete the mission despite ill-translated information and concerns about O-rings (Center for the Advancement of Digital Scholarship, 2024). Seven people died as a result.

From a humanistic perspective, it's unhealthy to stereotype managers and supervisors according to their perceived leadership style, for instance, the 'Hard Nut Production Manager' or the 'Wimpy Operations Safety Manager', because the underlying beliefs and values that make up an individual's psychology aren't visible to others. This was illustrated by Sigmund Freud's Iceberg Theory showing the conscious and unconscious mind (Freud, 1915; Mcleod, 2024). Behaviours are visible above the waterline, but beliefs and values are below the waterline and not observable. The 'Hard Nut' may well have strong people-centred values and be an OSH Champion, balanced with a pragmatic approach to risk acceptability. The 'Wimp' may be a quietly spoken strategist who takes a longer-term view on building capability through consultation.

It's worth noting that sub-conscious stereotyping can act against introverts compared to extraverts. In operations, this becomes problematic when popularity is confused with effectiveness. According to Lieg and Furtner (2023):

> Extraversion and its facets of assertiveness and sociability were identified as stable predictors for leader emergence and effectiveness. However, recent research suggested that extraversion may lie in the eyes of the beholder; it might not be the leader's possession but their followers' attribution of the trait that shapes these criteria of leader success.

(Lieg and Furtner, 2023)

The following comments were made about introverted leaders:

> our study shows that while the more effective leadership styles indicated by the full range leadership model (Bass, 1985) might be more suitable for extraverted individuals to engage in, effectiveness should not be equated with extraversion. At least one of the effective transformational leadership styles, intellectual stimulation, was perceived as more characteristic for an introverted personality. This finding is in line with recent literature on the positive influence of introverted leaders on proactive teams…. (Grant et al., 2011).

(Lieg and Furtner, 2023)

There is a key takeaway from this and other studies from a practical perspective.

*Individuals shouldn't pigeonhole themselves into being a certain
type of leader or blindly follow a leadership mantra.*

*They should be the best possible version of themselves and develop a deep
understanding to manage their own risks and opportunities
as a leader.*

Leaders tend to have strong skills in self-awareness such as recognising their own risks and opportunities. The factors to consider from an individual perspective include: What are my physical residual risks (limitations)? What are my psychosociological residual risks (negative tendencies, vulnerabilities)? These are characteristics that are difficult to change in the short-term so the Four-Step Risk Management Strategy would suggest managing these residual risks then investing longer-term in personal development. Leadership can also degrade. An important part of self-management is to build self-awareness so that warning signs are recognised, and feedback is listened to so corrective action is taken. This may entail getting help and implementing some form of maintenance programme, for instance, having a mentor or coach following a planned programme. Organisations seldom if ever consider leadership capability from a risk management perspective along with groupthink that would reinforce that all is well at the senior levels of the enterprise.

PRACTICAL RISK LEADERSHIP AND DECISION-MAKING

An organisation's senior management must be committed to building risk leadership capacity starting with their own competencies. The same key themes have emerged in industry from numerous historic disasters. Strong management commitment and leadership have been repeatedly shown to be critical to the prevention of catastrophic losses and business blunders, as well as pre-determinants of successful organisations.

Gigerenzer (2015) wrote extensively about risk management competencies. The analysis started with questioning if people are stupid, and it was suggested that society is largely risk illiterate. It was observed that the underlying obstacles to skill development include the way information is presented, for instance, statistically without context. Emotion also distorts interpretation so that circumstances such as COVID-19 that elicit a collective anxiety response, gain more attention than day-to-day facts such as road statistics, which are significantly worse. There's a need in industry (and society more broadly) to improve risk literacy as a life skill.

There are three ways to improve learning which can be derived from Gigerenzer's work:

1. Present information in a digestible form to deal with risk and uncertainty (with the inference being to tailor this information so that it's audience specific);

2. Experts are part of the problem because they make the same mistakes as members of the general public (with the inference being don't rely solely on 'experts');
3. Complex problems don't require complex solutions (with the inference being that simplicity is good).

What tends to happen is that an illusion of certainty arises. The following quote speak volumes:

> Trying to apply standard risk tools to uncertainty … can do tremendous damage, yet some experts act like problems of uncertainty are problems of risk so that they can use their standard mathematical tools rather than facing the real world.

> *(Gigerenzer, 2015)*

Risk leadership starts with understanding the operational reality of the organisation.

A different approach to running operational risk assessment sessions will be discussed later. For the moment, the discussion centres on building this operational understanding. What tends to happen in business is that senior managers find it difficult to schedule time in the field because there are constant demands. An organisation that is serious about building a risk- and opportunity culture underpinned by strong leadership will ensure that time in operations is a mandatory KPI for high-level decision-makers; however, this is often poorly executed.

From a site perspective, the worst way to go about this is for a contingent of senior managers in new site uniforms and shiny steel-capped boots to arrive in the workplace like a VIP visit. There are several reasons for this. Firstly, operations will prepare and ensure that everything looks good.

During the site visit, the senior management (VIP) party will receive information that is visually and verbally filtered compared to day-to-day operations, leaving them with a sanitised perception of the core business environment.

Secondly, operational personnel will likely perceive a distinct 'them and us'. They may feel like the visiting party is motivated by assurance rather than wanting to understand their operational reality. Thirdly, the party will likely bring their groupthink with them. The danger is that the field trip can result in a 'feel good' visit for senior management that operations endure until they can get back to day-to-day business. The better approach is for individuals or pairs to arrive with minimal prior notification and with an open demeanour, visiting key areas of the site but also showing interest in the work of contractors and special projects.

There's compelling research over many years, demonstrating that senior leadership is the key to high OSH performance. Companies with the lowest lost-time injury rates are those with the highest level of management commitment and employee involvement (Most, 1999). These high-performing enterprises have several traits.

Firstly, OSH efforts are equal to other organisational concerns and are fully integrated. Secondly, their cultural characteristics include specific management behaviours which improve safety and performance such as treating employees with respect, providing positive feedback, and encouraging suggestions (Erickson, 1994). Whilst the above speaks about OSH, the same traits apply to leadership of production performance, quality management and other functional areas of an enterprise. In this regard, it's difficult to fathom the concept of leadership being discipline specific, for instance, 'safety leadership'.

If an organisation is chronically immature in risk management and workers are being injured, then a specific safety strategy may be required to counterbalance production-centricity. Under these conditions, advocacy of 'safety leadership' may work to correct a downward and dangerous organisational path. In businesses that have embedded inroads into OSH compliance, discipline-specific branding may be counter-productive creating silos, wasteful overlaps between functions (safety, environment, quality) and obstacles to improvement.

There's no such thing as 'Safety Leadership'. There's only 'Leadership'.

One of the core competencies is the ability to make informed risk and opportunity decisions as a leader.

In *PSM1*, six broad competencies were identified based on the work of Most (1999). These still hold true for contemporary leadership underpinned by risk and opportunity driven capabilities:

Competency 1: The ability to think in terms of systems and knowing how to lead systems.
Competency 2: The ability to understand variability of work in planning and problem solving.
Competency 3: Understanding how we learn, develop and improve, leading to true learning and improvement.
Competency 4: Understanding people and why they behave as they do.
Competency 5: Understanding the interaction and interdependence between systems, variability, learning and human behavior; knowing how each affect each other.
Competency 6: Giving vision, meaning, direction and focus to the organization.

(Most, 1999)

The ability to think in terms of systems is fundamental to effective risk and opportunity management and this entails understanding how the work is done. In steady-state operations or long-term projects, a leader's role is to champion continual improvement drawing on the depth and breadth of competencies of the workforce. They must have the capacity to enable individuals and teams to achieve their higher potential. Transformational leadership, whilst it has been written about for decades (Bass et al., 1996), has renewed interest as a means of achieving capacity over the longer-term.

Helmold (2021) discusses 'Transformation Leadership in New Work' as a chapter in *New Work, Transformational and Virtual Leadership. Management for*

Professionals and presents a triangle of what management can't demand. This includes passion, creativity and self-initiative (illustrated in the article at the top of the triangle). They can, however, elicit intellect, hard work and obedience (illustrated in the reference at the bottom of the triangle). New Work is characterised by both job enrichment and job empowerment. The former involves raising personnel satisfaction by providing more tasks which deliver increased responsibility, autonomy and decision-making thereby improving the quality of the job. Job empowerment describes strategies for increased autonomy and self-determination so personnel can represent their interests independently (Helmold, 2021).

Transformational leadership requires managers and supervisors to loosen their grip on control without relinquishing it. Their role is to provide a guiding hand and empower their subordinates.

The words above are easily stated by the executive, but managers and supervisors need more than words. They need assurance. In the backs of their mind, they'll continue to believe that if something goes wrong, they'll be held accountability. Often the organisation's history will have examples of managers and supervisors being blamed for losses. These bad experiences tend to remain deeply entrenched in the minds of those still within the enterprise (the survivors) which encourages face-value commitment to change initiatives rather than heartfelt resolution.

The organisation therefore needs a clearly articulated and consistent method of influencing strategic and day-to-day operational decision-making processes and outcomes. In *PSM1*, the 'Reasonableness Test' was developed to illustrate the process underpinned by the organisation's values and a "Values Statement". Over the last 20 years (since *PSM1* was published), many mature, modern enterprises have developed and embedded their core business values. It's a mistake to assume that these values (like the policy of 'Safety First') will consistently translate into operational leadership behaviours in the presence of production-driven performance pressures.

Instead of simply relying on values such as safety and respect, everyone in the organisation must come to understand the organisation's Risk Ethos – the overarching principle by which decisions are made. This is particularly necessary for leaders to be confident that they will receive the consistent and fair support of senior management. Any negative legacy issues must be left in the past. An example of a Risk Ethos could be:

We, [insert organisation name], manage risk to achieve safety, production, and quality concurrently. Every decision we make is for safe production and a quality job. This is our Risk Ethos. This is how we work.

The purpose of the Reasonableness Test is to implement the Risk Ethos for informed risk- and opportunity-based decisions. A key principle of this approach is that decisions made using this process stand up to the scrutiny of the management levels above the decision-maker. They too must commit to and work to the Risk Ethos. This should be widely communicated and practiced in the organisation. The touchpoints that can influence the written culture of the enterprise should be identified, for instance, job descriptions and performance appraisals, so that the Risk Ethos is integrated into these. Verbal communication processes should also be an opportunity to articulate and apply the process, which is described in Figures 8.2 and 8.3. These are for strategic and operational decisions, respectively.

Figure 8.2 illustrates the 'Reasonableness Test' and 'Internal Strategic Alignment' to achieve risk-based decisions that lead to the appropriate level of physical, financial, and human capital. On the top-left, the external environment is identified as an input to the decision-making process. Here changes may occur to governing and influencing forces that require the attention of the organisation, such as legislative amendments or changes in client demands. The internal environment also needs to be considered. Key factors include information about relevant systems of production and organisational capacity. For example, does the enterprise have the internal capability to deliver a new product line or to meet new statutory obligations for environmental emissions? Information about losses and risk metrics may provide an indication of vulnerabilities that could inhibit any organisational changes. These will inform management of existing residual risks and degraded states. New legislation may provide the trigger for replacement or significant modifications to existing infrastructure or equipment.

The collated external and internal information should undergo quantitative and qualitative analysis. From an OSH perspective, this aligns to the concept of

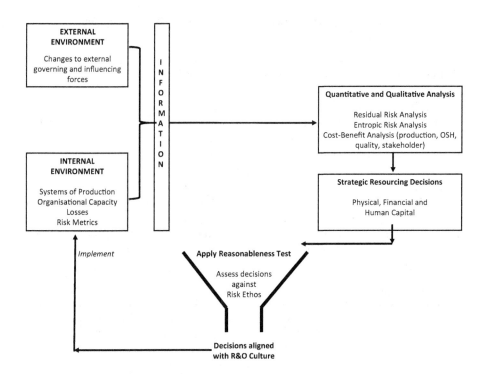

FIGURE 8.2   The reasonableness test applied to internal strategic alignment

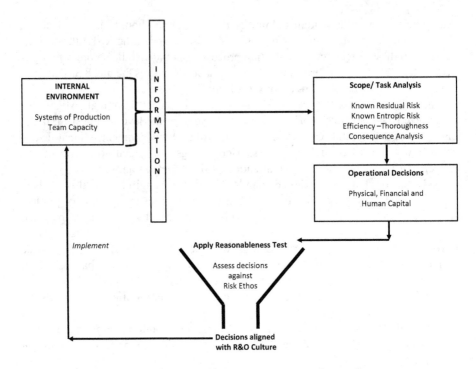

FIGURE 8.3   The reasonableness test applied to operational risk and opportunity decisions

'reasonably practicable' in terms of compliance; however, nothing in legislation states that production, quality and stakeholder needs can't be included in the cost-benefit analysis.

The figure shows that strategic resourcing decisions are an output of the information and risk analysis. There is a final test for preferred decision choices, and this involves assessing the fit with the Risk Ethos. Decisions that pass this test are aligned with the desired organisational culture. These decisions can be implemented, resulting in an action loop back to the internal environment where the required changes are made.

In the case studies described earlier, one of the most concerning phenomena at the highest levels of the organisations involved was the gradual degradation of the culture because of sub-standard decision-making processes. In the BP Refinery case, numerous corporate decisions were made that directly affected the operation's capacity to manage risk. These included failing to provide additional staffing during times of unit startup and shutdown.

> If staffing could be reduced 31 percent, reaching the 'Best class Solomon manpower index', the company would save $42 to $61 million. Such changes would result in staffing for normal operations only, where each operations crew is staffed at a 'lean' level, ie no additional personnel on hand to guard against 'peaks' such as safe-offs, absences and so on.

(p. 88, US CSHIB, 2007)

Would a decision such as this pass a Risk Ethos? Another root cause of the disaster was budget and staff reductions that impaired the delivery of training. At the time, BP used external consultant data to establish its own performance metrics even though such consultants didn't benchmark any specific process safety-related metrics (p. 154, US CSHIB, 2007). The inference with this hindsight is that a Risk Ethos may assist organisations to sense-check the decisions about the types of performance evaluation they implement. Are they using the right metrics?

This approach may conflict with other existing management strategies currently being implemented in industry. For instance, it's worth noting that the primary objective of the 'lean' management philosophy is to create value for the end-user stakeholder. This is done through optimisation of resources including the elimination of wasted time, effort, or money (Helmond, 2021) What a risk-based approach will do is likely provide more conscientious consideration of the impact of decisions over time. 'Lean' could have its deficiencies because there's a difference between rationalisation to improve efficiencies for the foreseeable future and cutting too deeply into the organisation's capacity. Severe cutbacks can limit the business' ability to adapt to and absorb the impact of external threats or to rebound from significant operational losses, such as a multiple workplace fatalities event. Unfortunately, these are too often the findings of major incident investigations. Cost-cutting carries risks and these should be understood by decision-makers with a view to not just the present, but also the horizon.

Figure 8.3 illustrates the 'Reasonableness Test' applied to operational risk- and opportunity-based decisions. At the site level, the internal environment is considered in terms of available systems of production and team capacity. The information is collated and analysed to identify known residual and entropic risks. In this part of the process, trade-offs and challenges become part of the discussion.

A factor that should be but is seldom considered in operational decisions, is 'Efficiency – Thoroughness Consequence Analysis'. If we do it faster, what are the consequences? If we do it more thoroughly, what are the consequences?

A mine production manager's advice to the maintenance manager that they need greater truck availability for production can be a 'loaded' comment. Does this mean, do whatever needs to be done (the minimum) to get availability today? Does it mean, ensure that the repairs and maintenance are done to a quality standard that will improve availability over coming months? There are likely different outcomes depending on how the maintenance manager interprets the priority and expectations, which will in turn affect the entropic risk level and potential for losses. Figure 8.3 shows that either way, operational decisions will determine the physical, financial, and human capital required to undertake the work.

Using the Reasonableness Test, each decision is assessed again the Risk Ethos. Would senior management want trucks available for today's production or, alternatively, equipment reliability to increase the likelihood of continuity of production, without failures and without consequences such as injuries or damages? In the 'organised chaos' of day-to-day operations, it can be difficult for the operational managers and supervisors to make 'reasonable' decisions; however, most carry a sense of accountability for their work. Most would appreciate assurance that if they make these choices based on a standard that aligns to overall company objectives, expectations, and the desired culture, they have the support of the boss and senior management.

If, after applying the Reasonableness Test, the mine production manager doesn't achieve today's production target because some trucks weren't available due to maintenance that manager has a justifiable explanation. Their objective was to manage the risk and pursue continuity of safe production (not simply today's production). If equipment availability is an ongoing issue, then this needs to be escalated to the strategic level of the company to consider options, such as resourcing of the maintenance department, supplies of spares, a revised preventative maintenance strategy etc.

A further reason for the Reasonableness Test is to address the issue of non-alignment. In Chapter 2, the Alignment Fallacy was explained. Numerous issues were identified that erode messaging from senior management as it cascades to the operational level. The executive may well support the Risk Ethos but without the tool to implement this as a consistently applied principle, operation's leaders may flounder with the tension of conflicting objectives. Senior management are saying they want 'X' but do they really mean 'Y'? People need clarity to make decisions confidently.

The Reasonableness Test is important because an organisation can put their managers and supervisors through extensive leadership training but, without criteria on which to make sound risk- and opportunity-based decisions, it's very difficult to transfer training into a complex, dynamic and demanding work environment. Supervisors deserve greater certainty and confidence when selecting actions with the knowledge that these actions are aligned with the system. Importantly, the workforce will observe the initiatives taken by the supervisor, who is more likely to earn credibility as a leader as a result.

All too often, frontline supervisors feel trapped between 'a rock and a hard place' when it comes to making decisions about safety versus production.

Decisions, both strategic and operational, should be assessed for alignment with the organisation's objectives and the desired culture. The test should also be applied to evaluate the appropriateness of current systems. For example, are there any practices that encourage shortcuts, introducing unacceptable risks such as engaging short-term contractors without basic competency verification processes? In the UK, during the period 2018/19 to 2022/23, 33% of fatal injuries were to self-employed workers even though they made up around 15% of the workforce (HSE UK, 2023) Overall, the fatal injury rate in this period was around three times that of the employee rate. Critical areas of the management system should pass the Reasonableness Test. An aim should be to correct areas of non-alignment to achieve congruence in organisational systems and the overall management ethos and direction.

The organisation's Risk Ethos is brought to the fore when supervisors have the authority and support to act in a manner that balances short-term gains and long-term benefits. In some cases, this may include shutting down a production line for safety reasons. If the team leader must justify their decision to a sceptical manager, then in future, they will likely hesitate and perhaps, ignore future safety issues to keep production going.

In the discussions so far, a few leadership considerations have been raised. Firstly, leadership isn't about personality such as introversion versus extroversion. Secondly, it's not always about high levels of participation and collaboration under all circumstances. Thirdly, stereotypes should be avoided because it's not possible to see an individual's internal values and beliefs. Finally, visible leadership involving VIP style site visits are readily prepared for to present operations at their best. The most appropriate style to adopt is according to situational variables. In *PSM1* for high-risk industries in particular, it was proposed that management demonstrate their commitment using an appropriate leadership style and decision-making process. Figure 8.4 (taken from *PSM1*) illustrates this concept.

The figure starts with management commitment and leadership cascading to organisational and operational resourcing decisions that create alignment; stakeholder needs are taken into consideration. The next layer down introduces situational risk variables. On the left-hand side, there are several risk factors that trigger an assertive leadership style. These include significant changes in the external environment, which may be threats or opportunities; however, these demand that the organisation act quickly. Due to time pressures and/or the need for expert knowledge,

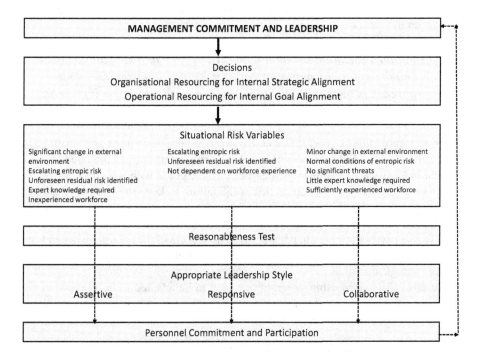

FIGURE 8.4   A top-down process flow of appropriate leadership at management and supervisory levels

collaboration isn't necessarily part of the process. This assertive style may also apply where the enterprise has an inexperienced workforce. The systems in place are likely focused on protection of workers rather than concurrently empowering them (however, this doesn't mean that care and concern is absent).

The mid-range situational risk variables are characterised by escalating entropic risk and/or the identification of an unforeseen residual risk and is independent of workforce experience. Under these conditions, management are responsive to any issues raised by operations and the workforce. The organisation's management elicit the input of those who are close to the risk source thereby providing themselves with many eyes and ears in the field. If the enterprise is an HRO, there is a chronic wariness of potential failure (as described further in Chapter 10).

The figure shows on the righthand side that in steady state operations characterised by minor changes in the external environment such as incremental legislative amendments, a collaborative approach is best taken. The risk characteristics of this situation include normal conditions of entropic risk and no significant threats. Expert knowledge isn't necessarily required and ideally, the workforce has sufficient experience to contribute to continual improvement of how risks are managed and how work is done to achieve optimal efficiency, safety and quality outcomes. The appropriate leadership style is collaborative.

Regardless of how decisions are made – whether using an assertive, responsive, collaborative leadership style – decisions are tested using the Reasonableness Test. Personnel are informed of the outcomes (who what when where and how) and the reasons that such decisions were taken (why), especially where these have direct impact on their work.

Information dissemination should be used to earn and sustain personnel commitment to the management of risks and opportunities. This transparency is important in situational leadership because it promotes trust.

Sharing information is good management practice because if there are information voids, there's the risk that these will be filled by rumour and covert dialogue. As was shown in Figure 8.1, organisational capacity can degrade with time and operationally, one of the factors that causes this is the failure to keep the workforce abreast of changes. Effective leaders understand that risk management isn't just about maintaining the standards of systems of production from a socio-technical perspective but also from a psychosociological perspective. The success of the enterprise depends on the small decisions that workers, supervisors and managers make each day, not just major strategic decisions.

In this regard, leadership programmes need to be adapted to build capacity for effective risk and opportunity management. There's a gap in how these are delivered especially under the auspices of 'Safety Leadership'.

Leadership programmes should be revisited to assess whether these deliver 'Leadership as Imagined' (LAI) or 'Leadership as Needed' (LAN).

OSH MANAGEMENT SYSTEMS AND LEGACY ISSUES

In Figure 8.1, it was identified that management systems can degrade. It's a professional trap for new OSH Managers to come into their role and make commitments to compliance because, ultimately, compliance isn't in their control even if they push for a culture of operational obedience. A call for compliance may be ill-founded if the management system is flawed. Some organisations, even those that are accredited to ISO 45001, can have gaps. This can result from failing to address key issues or promising too much. International Standards prescribe that processes must be in place to achieve accreditation but don't define the efficacy of those processes. The discussion about the ALARP assumption pointed out that mature operations can fall into a habit of prescribing further administrative controls after an incident even though this may add unnecessary complexity or cultural norms that drift towards a parent–child relationship between the system and workers. More and more rules are introduced. For instance, on some mine sites, drivers must keep their vehicle windows always closed and continue to wear their safety glasses (to avoid dust going in the eyes). A person who forgets this rule may be directed to pull over and told to correct the behaviour. Is this policing or care and concern? It can be difficult to differentiate in the field. Unfortunately, in the absence of a sound risk basis, much of the interpretation is left to the comfort level of the interpersonal relationship between the messenger and the receiver.

One of the first strategies of the new incumbent OSH Manager should be to ask operations if the OSH management system works for them (by inference, or against them) and to do a gaps analysis of the system. This should include a few key areas. The first is the difference between work as imagined and work as done. The second is any deficiencies in risk assessment with the view to identifying the justification for various tools. Some programmes are legislated, for instance, fatigue management, so these fall into the category of 'needs'. Any processes that don't add value from a risk management perspective should be scrutinised deeply. These are most likely 'wants'.

'Wants' can become pervasive with management's desire to be seen to do something. For instance, some corporations adopted behaviour-based safety (BBS) observations which entail a worker undertaking a visual review of their colleague's task performance. These are completed on the required form and sent to the OSH Department for review and entry into the system. Where the HR residual risk is high because of low competency levels within a high-risk environment, these can be beneficial by providing feedback to inexperienced workers on matters they might otherwise not be aware of. In an office, however, requiring project engineers to assess each other at their desk or making a cup of coffee, is a low-value, almost insulting, waste of time. If a programme doesn't have a sound risk-basis, why do it? What makes it worse is that this information is often input into the system to simply generate numbers. Resources are better spent on ensuring that safety critical controls for fatal risks are in place and are effective. These types of programmes that fall into the 'wants' category, tend to be implemented across all business units within an organisation rather than selectively on a risk-basis. The argument for this approach is that senior management wouldn't expect operations to do what they're not prepared to do themselves, but no one questions the value. The data goes into the system and for what purpose?

Operations carry the burden of ill-directed rules and cumbersome processes. There are several thought-leaders in the OSH profession who are questioning the rising power of the discipline and its legacy of bureaucracy, especially given that fatality and serious injury rates have flattened.

There is a clear need to get back to basics in OSH risk management and to address real operational issues. Procedural inefficiencies cause risk transfer by generating additional production pressure.

It makes the job of the operations manager or supervisor increasingly difficult, but the politics of safety branding make it worse by contending that leadership involves championing the mantra even though operationally, there may be a serious disconnection between aspiration and practicality.

The solution is to differentiate 'wants' from 'needs' and right-size the integrated, operational risk management system.

Once this has been achieved in consultation with the end-users of the system, and rolled out to the workforce, it's then reasonable to expect compliance to the system. A well-designed system should in fact, support the right behaviours by providing personnel with the information they need. For instance, this is a basic example. When doing the check on a site vehicle before taking it on a long remote journey, the checklist should include only the things that need to be inspected to ensure the vehicle is safe and reliable. If any of these issues, such as a tyre with low tread, are found, the checklist should prompt the worker to get this fixed before embarking. As a detailed example, if there won't be any mobile reception on the journey or at the destination, a statement should be on the checklist advising that a satellite phone can be obtained from the supplies department, who will check it's charged and operational. Checklists that are informative and practical are useful tools and difficult to argue against when designed to deliver OSH and efficiency concurrently. The objective of these tools should be to avoid losses – injuries, damage, equipment breakdowns, communication blackspots and time delays that have a flow-on effect on production. OSH professionals should facilitate the discussion about needs, but the solutions should come from operational managers and supervisors. Let them lead and be seen to lead.

The key message in this section has been to ensure operations have the tools they need to manage risk effectively. If this hasn't been reviewed, then it's risky to assume that the management system is working and that if something goes wrong, it will protect the business or its officers from statutory fines or litigation. If the system, after the gaps analysis, is found to be deficient, then a strategic plan is required to close the gaps. These should be prioritised on a risk basis. Initially, the system may be centralised to discipline experts who work in consultation with end-users, however, progressively, it should become increasingly decentralised to risk owners. For instance, historically, high-risk work systems such as permit-to-work have been developed and managed by the OSH Department. The rationale for this was perhaps that the OSH profession was perceived to be the expert in mitigating high risk safety issues. A risk-based approach entails ownership of the risk by those who are primarily affected and have the budget to manage the risk.

When a new project is planned, often the project managers will revert to tradition and expect the OSH Manager to be assigned this work, but this doesn't necessarily make sense. For instance, for construction of a major hazard facility, the highest risk in the project life that necessitates a permit-to-work system is when operations undergo shutdown maintenance. It's not until construction and commissioning have been completed that the plant is fully energised. The Hazardous Facility Maintenance Department therefore would be the primary user and should lead the design and implement of this critical control system, with other departments, being secondary users. An example of the latter is when tradespersons need to do trenching where there may be underground services on the site. A better risk management decision in the longer term of the facility may be to assign ownership of the permit-to-work system with the primary user and for the OSH Department to focus on verification of use and efficacy of the system (such as an audit function).

Whilst this discussion has diverted into management more than leadership, how the organisation implements its risk management strategy will determine the structures and accountabilities of risk leaders. Traditional approaches to OSH management have divested process ownership of many elements of the management system (because everything has a safety component) to the OSH Departement, whether this makes sense or not. For instance, why, in many organisations, does the OSH Department design pre-operational checklists for vehicles, plant and equipment when they are clearly not the users, the inspectors or the maintainers? Ownership of elements of the system should reside with the risk owner – the department who controls the resources to ensure that the system of production's residual risk is managed, and the entropic risks are prevented. This places the OSH profession clearly into a service role.

The role of the site OSH function is facilitation and verification of the integrated operational risk management system. The accountability for risk mitigation processes should reside with the Risk Owners.

This means that OSH professionals should facilitate risk management processes that have a significant safety component, such as management of change reviews. They can deliver high value by being the enabler bringing various operational functions together where there are overlaps to achieve simplification, standardisation, and reduction of complexity of practices and tools commonly used by different departments. Importantly, they should own the internal audit function involving verification of the efficacy of risk mitigation processes, especially those with fatal and serious injury or harm potential.

There are innumerable topics that could have been covered in this chapter about leadership. The focus has been on some of the key challenges that both the OSH profession and operational managers face. There are currently some serious legacy issues which have led to division in the OSH profession between those who advocate traditional safety (Safety I) versus those who prefer Safety II, 'Safety Differently' etc. From a practical perspective, both are needed.

Hale et al. provided definitions of the two approaches in 2012. These are useful because they get back to basics:

> Model 1 is the classic, rationalist view of rules as constraints on the behaviour of those at the sharp end of operations imposed by experts situated higher in the hierarchy and removed from the day-to-day work. Following rules is seen as the best way of behaving in all situations that can be envisaged, while rules constitute constraints on the errors and mistakes of fallible operators, who are portrayed as not having the time or competence to come up with those rules. This is an essentially top–down view of a static set of rules that should not be violated.
>
> Model 2 emerges from sociological and ethnographic studies and from management science, and focuses on predominantly complex, high-technology operations.... rules are socially constructed patterns of behaviour, deriving from experience with the diversity of reality. Rules are local, situated and have many exceptions and nuances. The experts in this paradigm are the operators, and this view is essentially bottom–up and dynamic. This model sees written rules as necessarily simplified abstractions of this complex reality, to be used as generic guidance, but adapted to each different situation in real life.
>
> *(Hale et al., 2012)*

From a practical implementation perspective, combinations are necessary to build leadership capability top-down and bottom-up. The two views aren't as binary as some commentators in the OSH profession suggest. Following are some of the main reasons for adopting both approaches concurrently from practical experience:

- Without senior leadership commitment, achieving buy-in at the operational level is problematic. If OSH is relegated by production, OSH risk management efforts will be eroded by operational reality and the frontline supervisor will be stuck in the tension between safety and production.
- Bottom-up risk management enables the workforce to own operational deliverables for safety, production, and quality. As the organisation matures, the OSH management system can progressively become more integrated and decentralised to enable day-to-day development of risk leadership capability by participating in practical, risk-based activities.
- Bottom-up is required to ensure that procedures and operational practices reflect 'The Work'.
- Top-down is unavoidable in large organisations that have a corporate head office where standards are set and that require consistent data across operations for governance.
- The OSH management system should mature progressively to facilitate a multidisciplinary approach to risk management and take advantage of synergies across functional areas such as environment, engineering, quality, security and human resource management.
- Risk management at both the strategic and operational levels is about loss prevention. This can't be achieved by singular top-down or bottom-up strategies or without a feedback loop.

- OSH information should be readily available to those who work to it, should be easy to read and follow, and should be improved if insufficiently prescriptive in high-risk critical areas or overly prescriptive for low-risk tasks.

It's important the OSH Department have a clear understanding of their role within the organisation and one of the most poignant things to remember is that credibility is earned not just given because of power or authority. Credibility is the key to leadership.

SUMMARY

This chapter opened by presenting the Entropy Model for Organisational Capacity which shows that four factors that build capacity have residual risks and can suffer degradation, leading to organisational losses. It was explained that deterioration of leadership and culture particularly were root causes in many major disasters, including those presented in earlier chapters.

It's not possible to cover the depth and breadth of leadership strategies and challenges in a single chapter so the discussions have been narrowed down to focus on some of the key issues that are faced in operations. There's extensive literature available which presents 'safety leadership' as a panacea for organisational woes. When working in operations, facing dilemmas such as efficiency versus thoroughness, short-term versus longer-term needs, and resourcing constraints, it can be difficult to come to terms with multiple functions each advocating their contribution as the most important. 'Safety is our highest priority'. 'Our customers won't be happy if we don't deliver on quality and without them, we don't have a business. Quality comes first'. 'Everyone knows production is core business. No production, no revenue, no jobs'.

Throughout this book, there's been a deliberate shift away from 'safety-based thinking' to 'risk-based thinking' which means a change in mindset. This reframes and perhaps debunks some myths about culture and leadership. Through a risk management lens, command and control management isn't necessarily 'dictatorship'. It may be necessary because of the high risk, short-term nature of a project in which there's little time for Safety II or 'Safety Differently' (Model 2) strategies. That doesn't mean that personnel are just numbers. The opportunity is to develop robust practices to effectively manage high risks in construction and short-term work assignments and ensure that these humanistic values are heartfelt despite this tough and demanding environment.

In this chapter, numerous solutions were proposed to address leadership challenges. A continuum of leadership styles was presented from assertive to collaborative based on the level of risk and the urgency of the decision-making process. This lent weight to a tried and tested way of leading – situational leadership. This dates to the 1960s and is based on the work of Paul Hersey who proposed four styles involving telling, selling, participating, and delegating. Operations are dynamic environments and leaders need to be flexible to changing circumstances. Different styles work according to the circumstances and the audience. Management literature can drift into idealistic leadership traits and fortunately, more is being written about authenticity. In the workplace, particularly in the Western World where individualism is a dominant trait (as opposed to collectivism), leadership credibility must be earned. It's not just given because of authority. Leadership programmes should be revisited to assess whether these deliver 'Leadership as Imagined' (LAI) or 'Leadership as Needed' (LAN).

Leadership programmes should involve discussions about decision-making traps such as groupthink where factors come into play that push a group towards consensus even though this may be a sub-optimal outcome. Stereotypes should also be avoided such as the 'heartless' Construction Manager or the vocal shopfloor 'troublemaker' who regularly challenges the status quo. Ironically, some leadership training programmes involve assigning different roles to participants including the 'devil's advocate' whose role is to raise risks, difficulties and problems. This is considered one of the most powerful roles compared to those who are expected to convey creative ideas, optimism or feelings. It's worth considering that the vocal person may have the courage to speak up whilst the rest of the group, sharing the same concerns, may remain silent. The person may be a potential resource – a leader in waiting.

The discussions highlighted that 'Safety Leadership' can raise some serious challenges operationally. The supervisor must include a section on safety in the daily pre-start meeting to demonstrate 'safety leadership', which can come across as separate from the work. Some organisations require a 'safety moment' or 'safety share' at the beginning of each meeting and this can revert to safely crossing the road or wearing a seatbelt when a person is put on the spot to demonstrate their commitment to safety. It's as if the supervisor is expected to put on a 'safety hat' which is different from what they ordinarily wear. The OSH profession would do well to reconsider the value of 'safety leadership' and 'safety culture' to better understand the impact of these on the role of the frontline supervisor and their operational manager, with the view to safety being a subset of 'leadership'.

A key theme throughout this book has been the tension between safety and production, and the impact this has on decision-making. *PSM1* included the 'Reasonableness Test' which has been refined in this edition. The test involves using a Risk Ethos to verify the suitability of decision choices and to ensure this supports the desired organisational culture. Managers and supervisors need criteria on which to test decisions to ensure that short-term pressures don't lead to sub-optimal decisions. These can erode the quality of systems of production and the organisational culture over time.

A section of this chapter was dedicated to examples of how managers and supervisors can adapt their leadership style to situational variables and the current risk profile so that decisions are delivered transparently to build trust. The best way to develop competencies is through practical risk assessment exercises directly related to the work. By engaging the team in these processes, the supervisor earns credibility as a leader and helps to develop the KSAs of their subordinates. Unfortunately, current OSH management systems are paper-driven to the extent that practical approaches to risk management are often stifled. OSH professionals should explore ways to enable a pragmatic approach to risk assessment in the environment using the language of operations, then taking the information to plug it into the reporting systems that senior management require to understand the organisation's risk profile. Each operational risk assessment activity should be seen as an opportunity to delivery three outcomes. The first is the required output as defined by the system (e.g. the risk assessment), the second is the consultative learning experience for participants (building competencies), and the third is the opportunity for supervisors to lead (building capacity).

Finally, a challenge was put forward to OSH Managers to scrutinise the management system to differentiate 'needs' and 'wants' with the view to 'right-sizing' the system. A cautionary approach was suggested because it's a mistake to drive for compliance to the system if the system has gaps, has areas of non-alignment or isn't user-friendly. Honest discussions need to be held with operations to uncover whether the system is a help or a hindrance. Well-designed, practical tools such as checklists guide the desired behaviours. Poorly designed written tools create frustration and covert pushback.

An invitation was issued to leaders of the OSH profession to improve the delivery of the OSH function. It was proposed that the profession should focus on facilitation and verification of the integrated operational risk management system. Historically, too many aspects of the management system have been assigned to the OSH discipline simply because it's perceived that safety has touchpoints across so many organisational areas. In this regard, the profession has sometimes created a rod for its own back by accepting work that shouldn't have been theirs to own. Plant safety and operational risk assessments are an example.

In many organisations, there's lack of clarity on the role of the OSH function compared with risk owners. Basically, the OSH Department should be a service function to core business. Accountability for risk management resides with the relevant risk owner (e.g. plant reliability is a maintenance department accountability). To avoid confusion and potential conflict, OSH shouldn't be undertaking operation's work (such as conducting the pre-start inspections on vehicles) or divesting their work, for instance, advising business unit managers to write their own safety management plan for a new scope of work or project. Unfortunately, there are numerous OSH service delivery models in industry which add to the confusion. If this is an issue that is eroding the leadership credibility of the OSH function, then accountabilities versus responsibilities should be mapped objectively.

In the next chapter, the discussions will explore organisational capacity more deeply. Some of the topics covered include why people take risks as opposed to manage risks, as both an individual behavioural choice and an organisational cultural norm. The Capacity Reservoir which was presented in *PSM1* will be revisited in Chapter 9 and linked to the concept of resilience which has emerged as a key success factor for organisations.

REFERENCES

Bass, B.M., Avolio, B.J., & Atwater, L. (1996). The transformational and transactional leadership of men and women. *Applied Psychology: An International Review*, 45, 5–34.

Bass, Bernard M. (1985). *Leadership and performance beyond expectations*, London: Collier Macmillan, 481–484.

Center for the Advancement of Digital Scholarship, 2024, Thinking as a group – COMM 326: Small group discussion methods (pressbooks.pub).

Erickson, J.A., 1994, The Effect of Corporate Culture on Injury and Illness Rates within the Organization. Dissertation Abstracts International 55 (6), Doctoral Dissertation, University of Southern California, USA.

Freud, S., 1915, The unconscious. *SE*, 14, 159–204.

Gigerenzer, G., 2015, *Risk Savvy: How to Make Good Decisions*. Gigerenzer, Gerd: 9780143127109: Amazon.com: Books.

Grant, A. M., Gino, F., & Hofmann, D. A., 2011, Reversing the extraverted leadership advantage: the role of employee proactivity. *Academy of Management Journal* 54, 528–550. doi: 10.5465/amj.2011.61968043

Grieve, R., & Van der Stap, T., August 2020, Safety and entropy – a leadership issue. *Professional Safety Journal*, American Society of Safety Professionals, USA.

Hale, A.R., Borys, D., & Adams, M., 2012, Safety regulation: the lessons of workplace safety rule management for managing the regulatory burden. *Safety Science*, 71, 112–122.

Helmold, M., 2021, *New Work, Transformational and Virtual Leadership: Lessons from COVID-19 and...* Google Books.

Helmond, M., 2021, *Lean Management and New Work Concepts*. SpringerLink.

Hosier, F., 2015, Feb. 13, CBS chief: DuPont has safety culture problem. *Safety News Alert*. www.safetynewsalert.com/csb-chief-dupont-has-safety-culture-problem

HSE UK, 2023, Work-related fatal injuries in Great Britain 2023, fatalinjuries.pdf (hse.gov.uk).

Isiadinso, C., 2015, BP Texas City refinery disaster—Accident and prevention report. https://doi.org/10.13140/RG.2.1.2317.4569

Janis, I.L., 1971 GroupThink.pdf (pbworks.com).

Lieg, S., & Furtner, M.R., 2023, Introverted and yet effective? A faceted approach to the relationship between leadership and extraversion - PMC (nih.gov).

Mcleod, 2024, Freud's theory of the unconscious mind: the iceberg analogy (simplypsychology.org).

Morgan, P., 2006, *The Concept of Capacity*. European Centre for Development Policy Management.

Most, I.G., 1999, The Quality of the Workplace Organization and its Relationship to Employee Health. Abstracts of Work Stress and Health March 1999, Organization of Work in a Global Economy, American Psychological Association/National Institute for Occupational Safety and Health Joint Conference, Baltimore, USA, p. 179.

Tannenbaum, R., & Schmidt, W.H., 1973, How to choose a leadership pattern (hbr.org).

US Chemical Safety and Hazard Investigation Board, 2007, Investigation report – refinery explosion and fire, March 23 2005, https://www.csb.gov/file.aspx?DocumentId=5596 and update at BP America (Texas City) refinery explosion. CSB.

Van Popering-Verkerk, J., Molenveld, A., Duijn, M., van Leeuwen, C., & van Buuren, A., 2022, Sagepub.com. *A Framework for Governance Capacity: A Broad Perspective on Steering Efforts in Society*. Jitske van Popering-Verkerk, Astrid Molenveld, Michael Duijn, Corniel van Leeuwen, Arwin van Buuren, 2022 (sagepub.com).

9 Operational Capacity

RISK-TAKING VERSUS RISK-MANAGING BEHAVIOUR

People make risk- and opportunity-based decisions every day. An announcement comes over the radio about a traffic jam on the freeway and that crews are currently clearing debris from the road. A worker about to leave the office in the city has various options for the drive home. They take the freeway and hope for the best, or take an alternative route, or go to the bar where colleagues are having Friday afternoon drinks. They choose the latter and promise themselves they will only have one low alcohol beverage. Hours later...

What is occurring is a basic process of deciding to either take the risk or manage the risk of getting stuck on the freeway, given what is known about the situation at hand. In a personal context, the consequences tend to be self-centric. Convicted drink drivers for instance go through a complicated process reflecting on moral values. Retrospective accounts of their behaviour focus on four dimensions: drink-driving as non-voluntary behaviour, as a strategic behaviour, as a relationship they have with control, and with 'normalcy' (Fynbo and Jarvinen, 2011). Drink drivers display positive attitudes to their own ability to drink-drive with few adverse consequences (Stephens et al., 2017). Unfortunately, the risk takers in society tend to underplay any negative outcomes for themselves and seldom consider the potential impact on others. In cultures where heavy drinking is socially acceptable, these behaviours may be reinforced by normalisation of these deviant behaviours.

In the workplace, there is certain conduct that is simply forbidden such as consumption, supply and private sale of alcohol and other drugs. The system is designed to eradicate these unacceptable acts through policy, testing and disciplinary procedures and for minor alcohol offences, help through an employee assistance programme to deal with underlying issues. Consumption of alcohol and other drugs is in the extreme range of risk-taking at work. Other less apparent risk-taking choices occur as a normal part of task performance and collectively, reflects the culture of the organisation.

The relationship between behaviours and culture was presented as two alternatives – risk-taking versus risk-managing – in the Risk Behaviour Model (Van der Stap, 2008). It builds on Risk Homeostasis Theory (Wilde, 1982). Figure 9.1 illustrates this model. The first panel looks at the positive payoff of risk-taking when there's no negative consequence. In the second panel is the undesirable/unexpected outcome where an injury or incident occurs prompting the individual and the organisation to learn the hard way about the consequences of their actions.

The figure in the first panel starts the positive payoff of the risk-taking cycle with misdirected values or internal drivers (1) oriented towards short-term personal gain. This could include financial benefits and social acceptance. A person may believe that

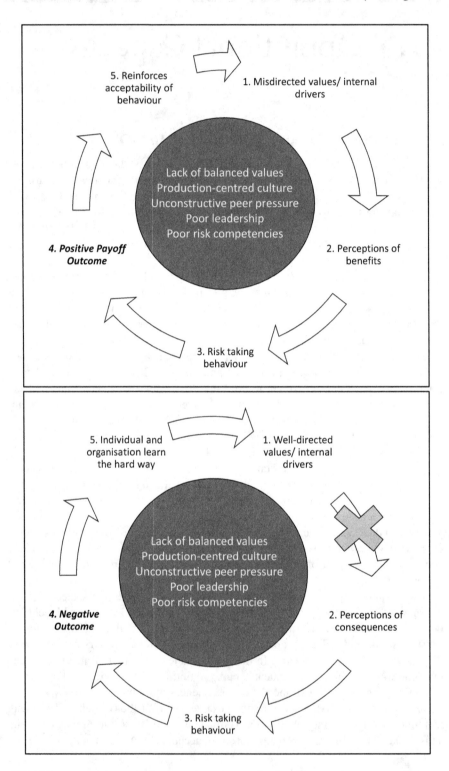

FIGURE 9.1   The risk-taking perspective with positive payoff versus learning the hard way

by getting the job done faster they can take on more paid tasks thereby increasing their earnings. They may 'just get on with it' to fit into a production-centred culture and to be accepted by their peers. These perceived benefits (2) lead to risk-taking behaviour (3) as shown. When there are no negative consequences, only a positive payoff (4), this reinforces the decision choice which tends to be repeated (5). The person's relationship with risk is consolidated. At the centre of this cycle is the collective impact. A culture driven by hard production targets and unconstructive peer pressure where few dare to challenge the status quo may be perpetuated. Consider the earlier example where underground workers falsified their vehicle checks, ignoring the faulty park brakes, to ensure that today's production wasn't hindered by lack of personnel transportation. The peer group may continue to focus on increasing rewards even more via collective effort to maximise production. As shown in the first panel, the perceived benefits of risk-taking are the primary driver of the actions that are taken.

Concurrently, as shown in the second panel, these directed values (1) are blocked so is serious consideration of the consequences (2). This mindset may be an individual or group trait that obstructs consideration of better (safer) alternatives. Risk-taking behaviour is acceptable (3) because personnel believe they will get away with it. The nature of risk reinforces this as in most cases, these deviances don't result in losses, simply due to the laws of probability. It's only where there's a negative outcome (4), such as a fatality or serious injury, that the individual and organisation learn the hard way (5). The centre of this side of the figure has the same destructive cultural traits.

This cycle is difficult to change unless the underlying contributing factors are addressed. When personnel don't have balanced values, they're not considering the risks of personal harm, injury to others or loss of future earning capacity that can result from serious injuries. Their timeframes are short – get it done and get paid. At an organisational level, the same patterns appear amongst the management team – today's production rather than consistency and continuity of production. There can be unconstructive peer pressure even in the supervisory ranks as shift bosses compete for the title of being 'champion of production'. The 'winner' has the crew who regularly achieves the most daily/ weekly/ monthly tons of ore, pounds of meat or widgets sent to market. What doesn't get considered is whether safety and maintenance issues are addressed on this supervisor's shift. The feedback from senior management may overlook the deficiencies in responsible risk management in favour of targets. It's only when there's a significant loss that poor leadership and risk competencies become evident if the executive is motivated to uncover these root causes. They may be reluctant to dig that deeply and prefer to level blame at site management to protect themselves from culpability.

In this reactive approach, depending on the severity of the injury, lessons may be short-lived as old habits re-emerge because the fundamentals of the culture, beliefs and values haven't genuinely shifted. In operations, this tends to be perpetuated unless there is a significant change in leadership direction from the top.

Operational managers will feel trapped with little authority to influence change if the source of the problem is at the executive level of the enterprise.

Experience teaches many professionals whose role has a strong connection with operational risk that organisations can fall into the categories of 'The Good', 'The Bad' and 'The Simply Irresponsible' (The author's colloquial version of Hudson's Maturity Model (See Foster and Hoult, 2013 reference for explanation). When starting a new role with a 'Good' organisation, the focus can be on continual improvement to raise performance further. 'The Bad' provide an opportunity for significant positive change which starts with some of the strategies described in earlier chapters. These include management system gaps analysis, facilitating activities so that managers and supervisors are actively involved in leading practical risk assessments, and holding discussions with senior management about areas of non-alignment, such as policy versus practice. Depending on the circumstances, it's not unreasonable to transform a poorly performing into a well-performing organisation within two years from a risk management perspective, provided that senior management has an appetite for loss prevention and exploitation of opportunities.

'The Simply Irresponsible' organisation is often characterised by a business in which senior managers feel little accountability to external parties. They may be business owners or have a high financial stake in the enterprise. Their positions may have been created by a relationship with those owners generating a sense of indebtedness that can compromise ethical decision-making. Serious OSH compliance issues may be ready identifiable to the trained eye but other anti-regulatory behaviours such as breaches of employment rights may also be occurring.

A spirit of non-compliance against legal requirements may epitomise the risk-taking profile of senior management which then cascades to poor treatment of workers apart from a core group considered to be within the 'inner sanctum'.

Referring to Figure 9.1, 'The Good' don't display the traits shown in the risk-taking perspective and have an aversion to learning the hard way from incidents. They maintain chronic wariness of failure and direct efforts to avoid it. 'The Bad' may be characterised by risk-taking at the start of a transformation process but are able to undergo a realistic self-assessment with well-directed guidance, to make the changes. 'The Simply Irresponsible' organisation won't break the cycle if the fundamentals of the business such as structure, ownership and senior management, remain unchanged. They'll continue to run the gauntlet of failures and potential consequences including prosecution. Direct contractors engaged by such enterprises risk also being exposed to such outcomes especially if their work is operational. They may compromise compliance to maintain their contracts.

As discussed earlier, historically, the response of the OSH profession for organisations needing significant improvement has included the introduction of behaviour-based safety (BBS) programmes and the 'Safety First' policy. This strategy assumes that the workers are the problem. 'Fix' the workers and the company's woes will magically disappear. Throughout this book, this approach has been thoroughly criticised. Operational managers, supervisors and workers need practical tools that assist them to manage operational risks (safety, efficiency, quality, reliability). When these elements of the system are in place, they tend to guide the desired behaviours.

Organisational capacity starts with having well-designed systems and tools for operations to use. Concurrently, organisations need to mature in terms of values, leadership, accountabilities, strategy, and self-discipline.

Many large corporations have undergone the process of defining their values. These often include terms such as safety, dignity, respect, accountability, courage, curiosity, community benefit and diversity. An internet search of various organisational websites shows that these values have proliferated in the last 20 years as part of defining corporate and enterprise identity.

In *PSM1*, a different set of values was described which align to risk management as a life skill. These were the sanctity of life (protection of safety) and the right to quality of life (protection of health). OSH is founded on these values. Alone, however, they don't provide balance or meaning in an organisational context where profit or service delivery (non-profit) also need to be considered. Good organisations / industries / economies create opportunities for long-term financial security and social acceptance. A worker can't sustain earnings for their working life if they suffer a serious injury or industrial illness. This value is needed to balance short-term versus long-term financial gains. Without conversations about these considerations, personnel don't have the opportunity to internalise what it means to manage risk as a life skill.

The final value proposed to underpin the Risk Ethos is social acceptance. The culture of the organisation determines whether risk-taking is accepted (socially endorsed) or opposed (socially intolerable). How this is conveyed will have numerous touchpoints including the design of the management system such as in training packages, in signs and symbols, and in formal verbal communications like planned team meetings. Messaging is also conveyed verbally through informal communications.

The four values – sanctity of life, quality of life, financial security and social acceptance – should underpin the enterprise's Risk Ethos. A decision to take risks that aren't endorsed is one that threatens all four values. For instance, if a production supervisor decides to keep the production line running until the end of shift to achieve their daily target, despite known equipment defects, there needs to be a correction factor identifying such actions as culturally unacceptable. In current management systems, there are no criteria to flag sound from unsound decisions that have an impact on risk levels.

A production-centred, risk-taking culture won't change without a mindset transformation at the senior management level which then needs to be cascaded to operations to at least the supervisory level. Figure 9.2 illustrates a more mature individual- and enterprise-level relationship with risk with well-directed, balanced values and internal drivers (in the first panel at 1). Benefits and consequences are weighed up (2) meaning that informed risk-based decisions are made. Timeframes shift outwardly to consider flow-on effects. For example, the production supervisor's target isn't today's production on their shift but continuity of production over a period that allows for contingencies across all shifts. This takes into consideration the work of all the production shift supervisors. Risk managing behaviour that involves ensuring critical safety and maintenance issues are addressed, are chosen (3). In this longer period which has been agreed with management (e.g. a month), the objective is

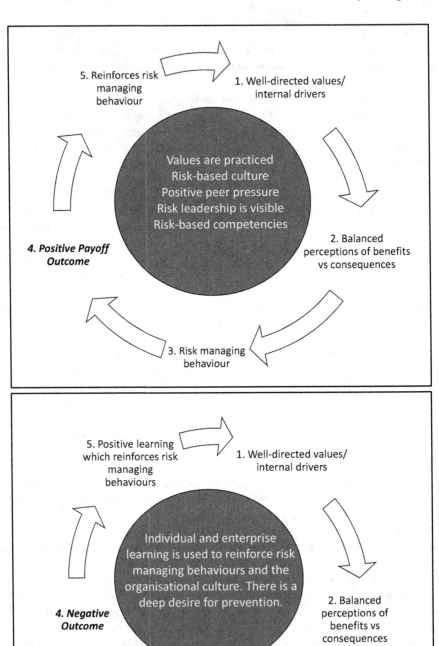

FIGURE 9.2 The risk-managing perspective with positive payoff and the learning process

continuity of production without disruptive losses. The positive payoff (4) is loss prevention, teamwork and reliability in the production process. This reward cycle reinforces this behavioural choice (5).

At the centre of the first panel of the figure, values are put into practice. A risk-based culture develops in which there's positive influence amongst the peer group to uphold the standards they've set for themselves. Risk leadership becomes visible and enables competencies to be developed where the risks must be managed operationally.

In the second panel of the figure, the cycle is repeated, however, if a significant loss occurs (the negative outcome at 4), the organisation and individuals respond with greater commitment to managing risk effectively. There's an appetite for learning from mistakes that reinforces the need to do better and a deep desire for prevention of losses.

From an operational perspective, a cautionary word is needed about production targets. If an unrealistic monthly target has been set, then operations are highly likely to fail. The further they slip behind, the more production pressure will mount resulting in extreme demands being put on the systems of production and the culture of the site. Consideration should be given to the impact of this on degradation and rising entropic risk levels. The Entropy Model highlighted the potential for losses under these conditions including incidents and injuries. If realistic monthly targets have been set but operations still fall behind, then questions should be asked about incremental losses that may be occurring along the entropic risk curve. It may be that systems of production aren't performing to management's expectations and there should be identifiable reasons for this, for instance, equipment downtime, employee absences, inability of contractors to deliver their scope on time etc.

Considering the key messages in Figures 9.1 and 9.2, how often do organisations provide a safe environment for conversations about workers' personal relationships with risk? Everyone has parameters around what they would or wouldn't do and under what circumstances. From experience, it's interesting to find out what activities people enjoy outside of work. In young males in particular, the profile often includes riding motorbikes, water sports, partying etc. They have a relationship with risk often with a positive mental association, but they may also have a relationship with safety that may have negative associations, such as when they were injured. Shared conversations are important as part of the learning and maturing process.

Talking about risk tends to be more engaging than always talking about safety. There are many stereotypes about safety versus risk. Safety is often associated with 'playing it safe' which is contradictory to society's entertainment where the hero is revered. Risk is synonymous with adventure, courage and excitement so it's little wonder that 'safety meetings' become uninspiring. Media such as extreme sports videos can be a captivating dialogue starter, especially when compared with traditional safety topics. In these conversations, people should be invited to consider their relationship with risk to establish regular internal dialogue. The objective is that future behavioural choices are made based on an informed relationship.

The organisation's objective should be that workers have regular internal and external dialogue asking, 'Is it worth the risk?'

Workers expect financial rewards and positive relationships from work and their primary reason for employment is to provide for life outside of work. Unfortunately, workplace injuries and illnesses don't 'clock off' at the end of the shift or on holidays or at retirement. Managers and supervisors should regularly ask, 'Is it worth the risk?' to convey chronic wariness of operational residual and entropic risks, and to also demonstrate care and concern for people's health, safety and wellbeing.

Real change occurs when the Risk Ethos is accepted based on the perceived compatibility between this philosophy and workers' own values. The probability that an action will be taken, or a behaviour will be followed increases dramatically when people develop a belief in its value (Topf, 2000). This means that the experiential learning provided to enlist employee commitment has to have the same rationale that applies to the management team. Some mature corporations have multi-layered programmes according to job level, but everyone (managers, workers, contractors) must complete the basic values-based training programme. This makes sense as there must be consistent messaging and language at hierarchical levels to achieve a values-driven culture. These programmes, however, shouldn't be the end of the learning process, otherwise, they become a 'tick-and-flick' exercise that puts everyone through a branding exercise. Personnel need opportunities to test the truth – to see if what they've learned makes sense and works for them in practice. In this stage, the supervisor is the most influential person affecting the operational behavioural choices of their team members.

These values-based training sessions need to be adapted to the culture of the society in which personnel belong. For instance, programmes that end in each participant standing up in front of their peers and making a verbal pledge to work safely don't fit an Australian audience. Other nationalities may be comfortable with the approach. Some cultures may prefer a collective commitment to an openly individualised undertaking. Overall, compliance-driven expectations shouldn't be part of the programme because they put people into a parent–child relationship with the organisational culture which misses the point of developing risk management as a life skill (not simply an organisational directive).

DAY-TO-DAY LEADERSHIP FOR SUPERVISORS

Operational personnel generally prefer hands-on learning over the classroom. With this in mind, the Entropy Model has been adapted to describe how leaders can engage personnel more effectively through interactions that develop new technical competencies concurrently building individual confidence and resilience. This is achieved through recognition and feedback that collectively enhances organisational capacity, with all interactions occurring in the workplace (Grieve and Van der Stap, 2020).

Figure 2.2 illustrated the phases of entropic risk in the HR system of production. Personnel can go through degradation, corrective action, maintenance, optimal performance then degradation again. The new employee contributes to initial degradation because there's a gap between their KSAs and those needed by the new employer. This can, however, also occur with experienced personnel whenever changes occur. It's a human performance reality that with change, there's often a decline in specific skills to do new tasks and an initial lack of confidence when undertaking the work. A concerted effort is required by the organisation to proactively develop the requisite competencies. Eventually, the individual or group of workers will achieve optimal performance (safety, efficiency, and quality), as shown by the adapted Entropy Model.

The concern is that cultural degradation within the organisation can occur leading to individuals' loss of motivation and/or self-belief for various reasons. Operationally, this is frequently observed with change fatigue; the ongoing processes of successive initiatives without periods of stabilisation. Site managers and supervisors particularly feel the pressure to just make things happen, make them work. There can be an information gap between what is expected to be done and why the new initiative from senior management is required. This potentially leads to distrust and growing cynicism about management's motives especially if more and more is demanded from workers without the resources or support to succeed.

Some management literature talks about adopting the mantra of being a person who is learning as a way of adapting to the many organisational dynamics which are outside the control of the individual. In practice however, it's the enterprise's responsibility to ensure that operational managers and supervisors are provided with the structures and coaching they need to implement new programmes.

For frontline leaders to develop risk leadership competencies, they need opportunities to rationally test and verify the benefits of risk management programmes and processes. Expecting them to just accept change is unrealistic.

Most frontline supervisors are action-oriented not concept-oriented, so they're more inclined to develop risk leadership competencies in the workplace than in the classroom. Active involvement in risk-based activities is a far better way to learn and it creates an environment where people feel more at ease. The following practical example illustrates this point.

In Chapter 5 on technological risk, a new truck was purchased for the remote exploration team to eliminate the reliance on contracted logistics providers. This also was intended to improve efficiency and reduce manual handling (the truck had a flatbed tray which could be loaded onto and off the back using the truck-mounted crane). A risk assessment needed to be done on this new piece of equipment. The operational manager and site team were expected the OSH manager to run a risk workshop in the training room on site. Instead, the team were invited outside to where the truck was parked, and the OSH manager started asking operational questions. The first was how the spare tyre, which was mounted on a hydraulic arm at the back of the cab, was taken down for a tyre change. This had been retrofitted after the truck had been purchased. Through a demonstration, it was found that not even the physically strongest worker was able to lower the tyre by turning the extension bar placed in a socket (a simple cranking mechanism). After considerable effort, the tyre on the arm was extended 1.5 metres from the side of the truck. It was impossible to lower it or return it to position. The mechanism was inoperable. The first part of the risk assessment was simply to test how things worked (or in this case, didn't work).

An operational risk assessment on a new or existing piece of plant or equipment can be led by the supervisor by testing and talking through how things work or are supposed to work, and the energies involved. For instance, on a drill rig (depending on the type), there will likely be pressurised systems (hydraulic and air), mechanical energies (such as rotating drill head), chemical energies (hydrocarbons, drill muds, dust generation), electrical energies (batteries), sound energy (noise) etc. One of the best ways to capture this information is in photographs with short descriptions in a user-friendly format, for instance, Powerpoint that can be used in a training session or printed as a handout. The risk assessment therefore also serves the purpose of being a training tool. If it's a corporate requirement for the information to be captured in a spreadsheet or database, then the OSH professional should undertake this work as a separate task that doesn't demand the time of the operational team. A brief verification session can be done with them later if a sign-off process is required.

Frontline supervisors can lead risk assessments when these are done as practical exercises focused on how things work (energies) and how things can go wrong, rather than an academic exercise centred on populating a spreadsheet or database.

These hands-on exercises use an inquiry process which provides a learning experience for all participants. Not everyone will have the technical knowledge to contribute to the risk analysis and those who don't will learn from those that do. This approach follows the principles of inquiry-based learning. Pedaste et al. (2015) discussed learning in five phases: orientation, conceptualisation, investigation, conclusion and discussion. These are driven by scientifically oriented questions most suited to an academic environment. In operations, however, it's better just to go from orientations straight into investigation which involves exploration, experimentation, and data interpretation (e.g. review of the operator's manual). From here, conclusions can be reached which lead to discussion. Pedaste et al.'s variety of inquiry phases are useful for the risk professional when assisting operations in these reviews. Some of the key points around communication include debate, sharing individual questions, communicating new understandings, elaborating, and justifying explanations. These are action-oriented conversations that can deliver improvements in participants' risk management skills, whilst also providing the supervisor with an opportunity to lead in a context where they feel confident. As suggested earlier, the paperwork (spreadsheet, report etc.) can come later. These are outputs for the system and shouldn't constrain the process if there's a better way of doing the risk assessment with high levels of involvement.

Unfortunately, in the OSH profession, the bureaucracy of the system and the subservience of operations to the system, tends to stifle practical approaches to risk management. The system defines the 'square' and people are discouraged from thinking 'outside the square'.

Risk leadership competencies are developed through these experiences, which will affect how new situations that require practical problems are approached and solved in the future. Traditionally, these types of risk assessments have been developed by the OSH function through a narrow lens that office professionals may be comfortable with (such as the spreadsheet), but which don't work in the field. For instance, the inability to retract the tyre on the hydraulic arm of the truck described above, would have serious consequences if this had been discovered during a journey. The operator could have been stranded in a remote area on the side of the road. Help could have been hours away. The recovery process would have been a costly exercise.

Many organisations lack an ongoing, strategic approach to leadership development after the initial onboarding of personnel or when changes are introduced into the enterprise. This was illustrated by Grieve and Van der Stap (2020) in the article, "Safety and Entropy: A Leadership Issue". Figure 9.3 is an adaptation of one of the diagrams used to explain the degradation of human resources (to the individual level) after onboarding and because of changes to systems of production.

The figure presents five phases of HR development that the leader can influence to minimise HR entropic risk and optimise performance. It starts on the left-hand side with Phase 1. This applies to onboarding or any initiative requiring management of change (MOC). Essential information about the organisation or about the planned change is communicated. An attempt is often made to provide a definition of success by sharing organisational goals or the objectives of the change. This establishes a common language and clear expectations. Initial training for onboarding will involve orientations and inductions, whilst for management of change, it will likely be competency training on the new equipment or new process. Operational risk management practices are often integrated into these technical training programmes, pointing out the major hazards of the site and the work with required controls. In the figure, Phase 1 shows that the entropic risk (dashed line) isn't applicable. This is because the new incumbent hasn't started work yet. For MOC, work is yet to begin with the modified systems of production.

In Phase 2, new incumbents have completed their onboarding and begin work. They introduce degradation into the workplace (shown by the downward dashed line). The frontline leader needs to be conscious of the potential for rising entropic risk which would be occurring in the opposite direction to the degradation line i.e. upwards (not shown). Relationship-driven strategies are required. The same applies after the initial MOC process and people are working to new systems, processes or with equipment. Leadership behaviours in this second phase focus on providing the team with an explanation of why it's necessary to work as planned or why the change initiative is needed. The dialogue centres not only on technical aspects of the work but also on building confidence by exploring concerns, listening, soliciting ideas, and providing encouragement. It may be necessary to reinforce values and the culture of the business if workers become cynical due to change fatigue.

The objective for the supervisor is to get through Phase 3 using upgrade strategies and move towards Phase 4 where the upgraded state is sustained. In Phase 3, the leader proactively develops and maintains the team by listening, acknowledging past successes, and providing perspective on the progress made. Where problems

Phase 1	Phase 2	Phase 3	Phase 4	Phase 5	Phase N
Onboarding or Management of Change (Core elements)	Initial Degradation	Proactive Development (Maintenance)	Optimisation	Cultural Degradation	
		Upgrade	*Sustain upgraded state*	*Degrade*	*Upgrade*
	Degrade				
Essential information: who, what, when, where and how	Explain why the way we work is necessary	Listen	Acknowledge high levels of competence	Listen	Continue Phases 3 and 4 as needed
Shared definition of success	Explain why the changes are needed	Acknowledge past success	Support autonomy	Acknowledge high levels of contribution	If changes occur to any systems of production, return to Phase 1.
Common language	Explore concerns	Provide perspective	Establish as expert	Facilitate self-reliant problem solving	
Clear expectations	Listen	Facilitate collaborative problem-solving	Provide positive feedback	Give redirecting feedback	
Initial technical training	Solicit ideas to address issues	Give redirecting feedback	Offer developing feedback	Provide perspective	
Operational risk management practices	Encourage growth and skill development	Encourage	Encourage innovation	Encourage	
	Reinforce values and culture		Value contributions	A concerted effort to address complacency	
			Aversion to complacency		

FIGURE 9.3 Appropriate supervisory leadership strategies by phases of employment

arise, these are resolved collaboratively. With minor issues, feedback might provide redirection and encouragement.

At Phase 4, the high-performance team is competent in the new ways of working. This should be acknowledged, and support given for autonomy. In this phase, work standards are maintained by the team independently of the supervisor's physical presence. When this occurs, it provides confirmation of resilience within the business. Progressively, some individuals may become the local subject matter expert or 'go-to-person' for specialised tasks. The leader steps back to optimise the capacity of the team to deliver and provides developmental feedback, encouragement for innovation and appreciation of contributions. This consolidates competence and confidence.

Importantly, the leader encourages the team to have an aversion to complacency either directly or indirectly. For instance, in earlier chapters an example was given of a maintenance workshop overhaul which resulted in a place for everything and everything in its place (HSE, efficiency and quality in design). When the supervisor returned to the workplace after a period of absence and observed that high standards of housekeeping had been sustained, feedback was given to one of the team members. The reply was, 'It always looks like this.' That's a sign of success. Later, the leader thanked the team for their professionalism (probably in colloquial language). Interestingly, in this culture what tends to happen is that individuals will correct each other if someone takes a shortcut, effectively creating a productive, healthy peer group pressure that fits the system and lifts the standards of those who tend to be laissez-faire. New starters who enter this environment are raised to the expectations of the group in a short period of time because standards are set from the outset.

Phase 5 illustrates what may happen if the standards of the group decline or external factors start to impact morale. The leader aims to lift the team by listening, acknowledging high levels of contributions, and facilitating self-reliant problem-solving. The team may need redirecting feedback, a sounding board to provide perspective, and reinforcement to achieve high measures of safety, efficiency and quality. In this phase, there needs to be a concerted effort to address complacency and underlying factors should be investigated without blame. Any degradation needs to be tackled by returning to Phases 3 and 4, as shown in Phase N. If changes occur to any systems of production, an MOC process may be initiated so the team returns to Phase 1.

Phase 5 can be problematic because cultural degradation, especially in safety, can produce organisational risks.

> The organisational risk is that leaders who believe their work was done once the employee moves into Phase 4 now see any failure (i.e., an incident) as the employee's fault, which leads to discipline and retraining as the corrective actions of choice. Cultural degradation is not because of a lack of KSAs, but a lack of recognition and feedback to ensure that the KSAs are applied consistently.
>
> *(Grieve and Van der Stap, 2020)*

Unfortunately, many leaders think that their development work is done in Phase 4, and thereafter, their primary role is to manage the workload. This allows complacency to set in.

*Conversations between supervisor and worker revert to technical
(e.g. how a machine is working) and social (e.g. what happening in sport
on the weekend) topics. The need for the worker to be engaged in the work
process becomes forgotten and the supervisor regresses to being a manager
rather than a leader.*

The OSH management system and cultural branding can result in bland transactional relationships when statistics become the primary driver of language rather than risk leadership, workforce capacity and relationships.

> The traditional safety measure of OSHA recordable injury rate and experience modification rate drive leaders and employees to adopt the 'be safe' mentality. This mentality is demonstrated by statements such as 'as long as everyone goes home with all their fingers and toes'.

(Grieve and Van der Stap, 2020)

(NOTE: The Experience Modification Rate Calculator is a measure used in payroll classification to determine the workers' compensation insurance premium. This reflects the company's safety performance and risk management. Experience Modification Rate (EMR) Calculator – Savvy Calculator, 2024).

OSH metrics are important, but these should stay firmly within the OSH function and reports to senior management. Billboards that display number of days without a lost-time injury (LTI) convey wrong priorities putting 'the cart before the horse'. These results are an outcome of numerous factors which may include effective operational risk management or unfortunately, non-reporting of incidents or management of injured workers so that they don't lose a day/ shift of work to fit the LTI classification. The site's senior managers would want to be certain that the result is down to good management if they're expecting the workforce to see this as a success they also own.

The situational and transformation leadership styles described above may seem daunting for some managers and supervisors to adopt. It requires authenticity in their relationships with subordinates and a sound foundation of risk and opportunity management competencies. These skills can be developed with time through the practical risk evaluation activities described above. OSH professionals have an important role to play.

*OSH professionals should enable real work, real conversations, real
understanding of risk and real results. They should do the OSH work by
taking operational information and putting it into the OSH management
system to meet hierarchical reporting requirements. They should own this
interface between operations and senior management.*

Managers and supervisors should have minimal bureaucratic burden and be provided with systems that integrate OSH and operational risk management into day-to-day work. In addition to the workload, their focus should be on communications that empower the team. As a societal comparison, the cycle shown in Figure 9.3 has parallels to backing a sporting team. When the team goes into decline, supporters rally and acknowledge past successes (Phase 3). They reflect to provide perspective on current performance. During periods of high performance having achieved numerous wins, complacency is considered a threat (Phase 4). The discipline that goes into training and execution is sustained. The coach seeks out other leaders in waiting within the team to value their contribution and provide developmental opportunities. At a future point in time, supporters remain loyal despite a series of losses (Phase 5). They don't just change allegiance when their team is down. The coach (leader) takes the reigns again to raise morale and team performance. This is a mindset that could be helpful to operational managers and supervisors working through the cycles of HR entropic risk. Motivation starts with self-motivation with the leader looking inwardly before they turn their vision outwardly.

There are too few conversations about successes in operations that celebrate a job done without incidents, on time and on budget because the work was properly risk assessed and planned with the involvement of key personnel.

The following is an example. A major seismic survey had to be undertaken by a specialist contractor on a mine site. The overarching legislative requirement is that contractors comply with the mine operator's safety management system (SMS); however, given the contractor's high technical skills, imposing the mine system entirely would have been counter-productive to effective risk management.

The footprint of the work involved numerous hazardous interfaces such as the active mining area, ore and waste stockpiles, an explosives storage facility, vegetation stockpiles that would be used later for rehabilitation and internal roads with high levels of traffic. This meant that there were many internal stakeholders who needed to be consulted to manage the risks. There were a few key critical success factors in how the work was executed:

1. Trust: The contractor was treated as a partner in the execution of the work. The traditional principal-contractor (parent-child) approach, still commonly seen in industry, was replaced with a mindset of enabling the contractor to deliver.
2. Stakeholder Engagement: Representatives from all parties who would be impacted by the scope of work and whose activities could affect the seismic survey were brought into the room. This included managers and supervisors responsible for the active mining area, drilling and blasting, production scheduling, the contract owners and the contractor.
3. Risk Assessment: Stakeholders participated in a risk workshop that focused on how the work could be done safely, to the required quality and productively using a multi-disciplinary, multi-objective approach not just 'safety'. In addition, inspections were undertaken of the proposed work areas to identify potential issues such as access to the tops of stockpiles and the interface with active mining. The risks and required controls were captured in a scope of work risk register.

4. Planning and Communication: An Operational Risk Management Plan was developed for the scope of work which clearly explained the required risk mitigations and communication practices. Importantly, it also explained the boundaries of accountability between the mine holder, the contract owner and the contractor. The plan was reviewed and fine turned with the input of stakeholders.

5. Execution: Everyone worked to the plan. A critical piece was the agreement that the contractor would be given authority over the area in which there were working to a radius of 50 metres. This was the case for the survey footprint except in the active mining area, which needed to remain under control of Mine Production because of high risks including heavy mobile plant and drilling and blasting activities. This approach enabled the contractor to work their own SMS whilst interfaces outside their work area were managed by the mine operator and contract owner.

The feedback from the contractor after the completion of the project without incident, on time and on budget was that they appreciated that the client was proactive in establishing trust in the contractor. This enabled a collaborative approach to safety and helped to identify and manage the 'real risks'. This facilitated stakeholder engagement and provided a direct line of communication if any complications or changes arose during the execution of the work. The example highlights the need to focus on the work and who is responsible for what.

A useful concept is 'front-end loading' of the project or scope through consultation, risk assessment and planning of the work for success. Effective risk management ensures that safety, efficiency, and a quality job are outcomes.

CAPACITY IN RESERVE, RESILIENCE AND RESOURCEFULNESS

The broader objective of investing time and resources in the leadership skills of managers and supervisors is that they in turn, become skilled coaches of their team members and more effective when working with each other. There's extensive literature available on leadership development and some of the better resources include those that explain sound fundamentals that can be applied in any workplace and at home, such as Stephen Covey's "*7 Habits of Highly Effective People*" and "*Principle-Centered Leadership*".

In *PSM1*, the focus of leadership development was on building organisational capacity at all levels of the hierarchy and creating a mindset change towards the value of human resources. This can be achieved in practical ways. As discussed throughout this book, both bottom-up and top-down strategies are required concurrently. The OSH discipline can be used as an example of a grass roots approach. According to OSH legislation, organisations must manage risks to prevent injuries and harm to the health of workers. It's incumbent on OSH managers to ensure their employer has a programme to prevent fatalities, serious injuries, and debilitating illnesses. These programmes are usually referred to as fatal risk critical control pro-grammes or major hazard standards (which were introduced in previous chapters).

Many workplaces have commonalities in relation to these fatal risks, for instance, electricity, working at heights, hazardous substances, light vehicles (company cars), fitness for work, emergencies etc. Each of these will have controls that are critical for injury and incident prevention which means that there are numerous checks to be undertaken on a regular basis. Medical emergencies are a relatively straightforward example that applies to every workplace. The controls required include having an emergency management plan; having a current contact list of internal personnel and external support services to call if an emergency arises; having trained first-aid offi-cers; having and practising emergency evacuation drills; having first-aid equipment available and kits that are properly stocked etc. Putting together a verification process involves having a schedule for checking these critical controls on a planned basis and assigning this work. Given that a workplace may have numerous fatal risks there may be a significant workload of verifications to be undertaken. The most practical approach is to decentralise this work to key personnel in each workplace. They col-lect the information which is reported to the OSH Department who then collate the data and report findings formally to senior management. The traditional approach of centralising the data collection using OSH advisors is impractical and defeats the opportunity for operational ownership and involvement.

One approach to make sure the work gets done is to ask the workplace supervi-sor to tell someone to do the checks when required and report back to them. This becomes another dull task that workers must perform, often being an irritation as it takes them away from their 'real job'. A different and better approach is to build capacity. The supervisor, in collaboration with the OSH professional, may invite their team members to take on various roles. In this case, the role would be to champion the operational emergency management function in their work area. In return, the volunteer would be provided with first aid training, evacuation warden training, and time away from normal duties to do the required checks. These include that first aid officers have current training certificates, that emergency drills are conducted as planned, and that first aid kits are properly stocked etc. The planning of the criti-cal control programme needs to include coverage. Several workplaces in proximity may only need one or two champions, whereas a separate workplace may need its own representation. How this is structured will depend on the risk. For instance, in a high-rise building, there should be at least one emergency champion per floor with coverage for absences, as well as several first aid officers and evacuation wardens. Many workplaces take this approach but without considering the benefits from a cultural perspective.

Decentralising safety critical controls and the verification process builds capacity and a strong operational risk culture.

In *PSM1*, the concept of developing capacity was illustrated using the 'Capacity Reservoir' which holds the collective leadership skills and competencies of the workforce. This has been updated in Figure 9.4 and the descriptor originally used,

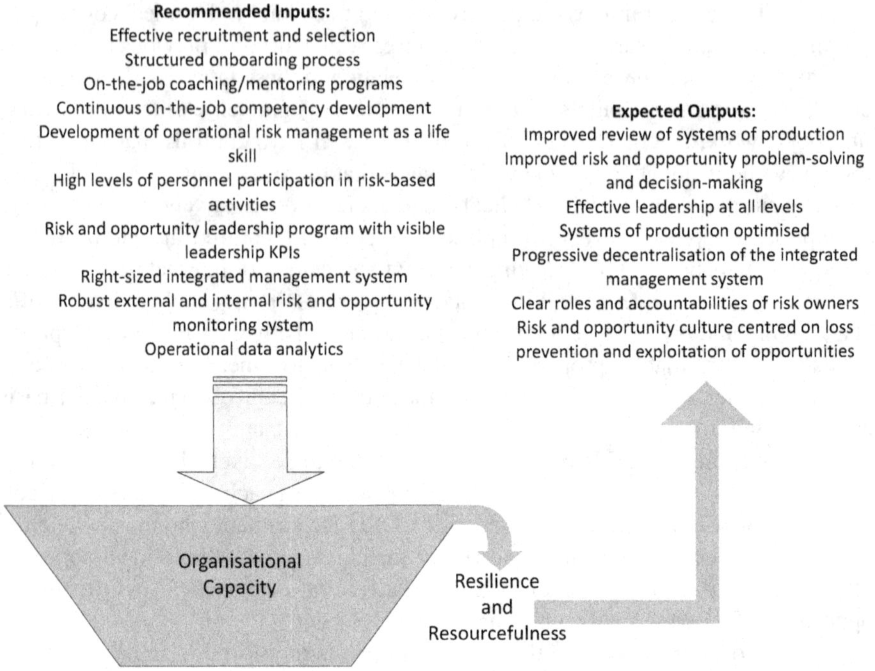

Recommended Inputs:
Effective recruitment and selection
Structured onboarding process
On-the-job coaching/mentoring programs
Continuous on-the-job competency development
Development of operational risk management as a life skill
High levels of personnel participation in risk-based activities
Risk and opportunity leadership program with visible leadership KPIs
Right-sized integrated management system
Robust external and internal risk and opportunity monitoring system
Operational data analytics

Expected Outputs:
Improved review of systems of production
Improved risk and opportunity problem-solving and decision-making
Effective leadership at all levels
Systems of production optimised
Progressive decentralisation of the integrated management system
Clear roles and accountabilities of risk owners
Risk and opportunity culture centred on loss prevention and exploitation of opportunities

Organisational Capacity

Resilience and Resourcefulness

FIGURE 9.4   The inputs and outputs of the capacity development process

'Resourcefulness' remains but 'Resilience' has been added to reflect the need for adaptation in the face of crises and significant threats beyond the direct control of the organisation.

The figure starts on the left-hand side with recommended inputs. The first is an effective recruitment and selection process to minimise HR residual risks. This is followed by a structured onboarding process to accelerate learning and mitigate HR entropic risks. After this, the critical process of providing on-the-job coaching or mentoring occurs. As described earlier, workers need support during the assimilation phase and whenever management of change is required. This is often missing in organisations and the supervisor-worker and organisation-worker relationship can become transactional after the initial enthusiasm of the new job has waned.

Importantly, operational risk management is identified as a life skill which has much more value than simply compliance to the system. This entails the capacity to identify risks within a dynamic environment, to problem-solve and make informed decisions, particularly, in a team situation. This competency is best developed through participation in risk-based activities which have a practical focus. This should become a natural experience for the supervisor who leads the team through the risk assessment considering safety, efficiency and the required quality of work concurrently. The discussion focuses on successful execution of the work. It should go without saying that 'successful' entails safe production to expected quality standards.

In Chapter 8, the discussions turned to visible leadership by board members, the executive and senior managers. It was proposed that these personnel should be given metrics for site visits to ensure they get an understanding of the operational reality of the enterprise they lead. Importantly, it's not about completing the site visit under ideal conditions but showing a deep interest in day-to-day challenges and opportunities. Through the eyes of a cynic, these visits should at least ensure that 'officers' of an enterprise have done their due diligence. In Western Australia for instance, an 'officer' is a person in the business who makes or participates in decision-making that affects the whole or a substantial part of the enterprise's activities (Department of Commerce, 2024). 'Due Diligence' entails taking reasonable steps to understand the nature of operations and of the hazards and risks associated with those operations (Worksafe WA, 2021).

As shown in the recommended inputs, capacity is also developed when the management system has been integrated (meeting the needs of multiple functions in single documents or tools) and right-sized to deliver what operations need to manage risk effectively. Management systems can either help or hinder organisational performance. These may be overly risk-averse resulting in safety process causing unnecessary bureaucracy and delays, which then result in increasing production pressures to complete work in less time. This was referred to earlier as low-value thoroughness for safety's sake without a sound risk basis. As an example, on a major construction project, one of the KPIs was the completion of a minimum number of Take-5s per person per day. (A Take-5 is a short personal risk assessment checklist usually in a booklet that the worker carries with them. A check is mandated for every time the worker enters a new work area and at other times according to the set criteria). There was about a thousand workers on site so the Take-5s were also in the thousands. Two people were employed full-time to enter the data into the system to generate the numbers, however, little was done with the data and the quality of the data was never questioned. This was a process that the construction managers and supervisors tolerated but didn't support.

The management system should focus on the major concerns – fatalities, serious injuries, and debilitating illnesses. The relationship between unsafe acts, near misses, minor accidents, serious accidents, and fatalities, which argues that if minor events are eliminated then fatalities and major injuries will be prevented, is flawed.

Heinrich's Triangle (Wikipedia, 2024) which was the basis of this relationship between unwanted events, has come under significant criticism recently from the OSH profession. All organisations should focus on critical controls as their primary preventative programme. 'Right-sizing' should entail bringing a realistic perspective to the data collected from operations and the responses this elicits. For instance, a rise in reported first-aid cases means little without context. Unfortunately, misinformed senior management may be unnecessarily concerned. The increase may be good news that the reporting culture has improved and hopefully, this isn't just for safety but also efficiency and quality. It could mean that the capacity of the organisation to identify and rectify issues has improved. Alternatively, it could mean that the supplies of safety gloves have run out and people are continuing to work without them, leading to minor hand injuries. KPIs used to evaluate operational performance need to be contextualised.

Capacity relies on knowledge and sustaining the currency of that knowledge. Robust external and internal risk and opportunity monitoring is required to protect the enterprise from any factors that can cause it to become vulnerable. This was particularly apparent in the earlier discussions about the Fourth Industrial Revolution and the rate of change that this has brought. At the most fundamental level, the enterprise must stay abreast of legislative changes, client demands and social trends that affect its operations. For instance, in an earlier chapter, it was mentioned that some large corporations are now requiring their contractors (including small operators) to sign-off on and be prepared to be audited against the corporation's modern slavery policy. Risk monitoring processes are needed in businesses to ensure, for instance, that missed or disregarded communication from the client doesn't result in loss of contract.

The final input shown is operational data analytics which are linked to these monitoring systems. The bottom line is for managers to better understand the strengths and weaknesses of their business and how these present risks and opportunities. In Chapters 3 to 6, each of the systems of production residual and entropic risks were explored and it was shown that different organisations can have quite distinct risk profiles according to the composition of these systems. Some businesses have higher levels of HSE in design in their technologies and infrastructure that significantly reduce their residual risk profile. Operational data should assist management to maintain the quality of systems of production to optimise efficiency, safety and quality.

Figure 9.4 suggests that these inputs collectively determine the enterprise's capacity (shown by the reservoir shape), the investment in which should lead to what is shown as an overflow of resilience and resourcefulness. This, like the transformation from 'safety-based thinking' to 'risk-based thinking', illustrates an important shift in mindset without using a slogan or branding such as, 'People are our greatest asset'. What it means from an operational as well as strategic perspective, is:

Managers and supervisors should view their workforce's capabilities for potential rather than finite competencies to fit the various job descriptions the enterprise has created.

The term 'resilience' has been around for more than 40 years but has become more significant in the last decade due to the rate of social and economic change as well as challenges such as COVID-19 which demanded adaptability. There's extensive literature available on this topic and a useful summary is provided by Martin-Breen and Anderies (2011). They state that resilience building involves increasing the capacity of individuals, communities or institutions to survive, adapt and advance despite acute crises and chronic stresses. This requires a multifaceted, interdisciplinary strategy and a systems view (Martin-Breen and Anderies, 2011). At a global level, the term 'wicked' problems has emerged to identify current inherent catastrophic conditions such as poverty and climate change. 'Wicked' problems within business largely remain 'undiscussables' due to habits that result in organisational silence, as discussed in previous chapters. Many large enterprises, however, talk about building resilience but still lack the courage to challenge issues of non-alignment such as silos, internal conflicts and the risks of leadership degradation.

Directing the term to an organisational focus, certain key correlations relate to systems. According to the researchers, these include 'responsiveness to regulatory feedback mechanisms'. Put simply, the system (both 'hard' and 'soft' components described earlier) should be built to include monitoring to ensure that these continue to perform as expected, and if not, this should trigger corrective action. From an

HSE in design perspective, technologies should contain 'modularity of interactions between parts, entailing failure containment' (p. 53, Martin-Breen and Anderies, 2011). If one part of the physical systems of production fails, this shouldn't set off a chain reaction that destroys the entire system. There should be redundancies to cover the contingency of component breakdowns. The final property of resilience is that the enterprise can internally self-organise to adapt.

One of the key themes of resilience is the ability to learn and adjust to change without being forced to change in an ill-directed way. In this regard, the term originally used in *PSM1* – 'resourcefulness' – shouldn't be disregarded. Resourcefulness is 'the ability to find quick and clever ways to overcome difficulties' (Cambridge Dictionary, 2024) which resonates well with how operations need to work. There's a positive relationship between managerial job performance and resourcefulness (Sahin et al., 2015). The foundations of this skill include affective, intellectual and action-oriented competencies. Affective behaviours include the ability to have a sense of equanimity and a problem-solving orientation, to delay immediate gratification, and importantly, to demonstrate proactive involvement in the face of challenges (Sahin et al., 2015). The intellectual aspect requires the ability to deal with information and to see the linkages and dependencies within the system. This leads to planning and evaluation of alternative decisions and likely outcomes. There's also an element of self-reflection, which indicates an appetite for personal continual improvement.

The final foundational skill is action-orientation. Importantly, this has two dimensions – task-related and people-related. The former fits traditional understandings of competent management including attention to detail and to timeframes. People-related competencies are described as involving the ability to establish empathic, non-directive, supportive and trusting relationships (Sahin et al., 2015). This relates well to the prevention of HR entropic risk through managers' and supervisors' support of workers in their engagement with the work, as described in Chapter 8.

In Figure 9.4, there are some clear expectations about the return on investment of developing this capacity. This includes the outcomes listed on the right-hand side of the figure starting with improved review of systems of production. The Entropy Model identified the need to clearly understand the sources of residual and entropic risk of systems of production that can lead to losses and potentially catastrophic events. With better review comes improved risk and opportunity-based problem-solving and decision-making, which was a key theme of Chapter 8. As the organisation matures, leadership capability should be enhanced at all levels – top-down and bottom-up. This includes greater participation and responsibility of Risk Champions in the workplace, particularly with frontline supervisors and workers contributing directly to loss prevention verification processes. As an example described earlier, each workplace should have champions for emergency response supporting the centralised assurance programme that takes operational data and presents it to senior management. The executive, in turn, should ensure required resources are provided for any corrective actions and cost-benefit driven, risk- and opportunity-based improvements.

The traditional role of managers has included clarifying roles and accountabilities. This is particularly important in risk management. OSH has touchpoints with many if not all areas of an organisation but that doesn't mean that the OSH function is accountable for safety performance. They don't control the resources for plant and equipment maintenance for instance, so can't be held accountability for technologi-cal failures. They're not in charge of the workforce so can only put systems in place for the prevention of alcohol and drug abuse such as testing regimes. Supervisors are the primary detection agents outside of this testing regime. The risk owner is the 'person or entity with the accountability and authority to manage a risk' (ISO, 2024). Unclear lines can create conflict which is wasteful and contrary to the traits of resourceful organisations.

The final expected outcome is a positive impact on the organisational culture. Cautiously, the label, 'risk and opportunity culture' is proposed, and this is a signifi-cant subset of the broader enterprise culture. In the same way that 'safety culture' and 'safety leadership' (discipline-based classifications) have been questioned throughout this book, another boundary should be avoided in favour of a clear understanding that that organisation's primary mindset operationally should be to prevent losses and exploit opportunities through safe, efficient, quality work that delivers on stakeholder needs.

In the next and final chapter, the discussions about capacity will continue. The principles of HROs will be explored to tie all contents of this edition together and to look at a way forward for organisations based on these principles.

SUMMARY

In this chapter, the dialogue about leadership that opened in Chapter 8 was contin-ued. The different choices that people make in the presence of risk were described using the Risk Behaviour Model. There were two opposites presented which were risk-taking versus risk-managing. These involved decision-making processes that result in what the individual considers an acceptable behaviour. This was largely influenced by the culture of the workplace and the outcomes. There are expectations that the chosen behaviour will lead to a positive pay-off and the avoidance of conse-quences. These behaviours perpetuate the culture and vice versa so both aspects need to be addressed to shift to a mature, risk- and opportunity-driven enterprise.

More broadly, businesses can fall into categories which were referred to as 'The Good', 'The Bad' and 'The Simply Irresponsible'. Different strategies need to be applied depending on the level of maturity of the enterprise and its senior personnel. Owners, executives and top managers who have a high-risk appe-tite, a production-centric mindset and short-time horizons are difficult to shift. Unfortunately, these can be destined for learning the hard way from incidents, inju-ries and losses. Some regard flouting the rules (not just in OSH) as a challenge. Those associated with such businesses like in-house risk advisors and contractors should seriously consider whether they can accept their risk exposure to such clients.

In contrast, many organisations have defined their values. These were discussed in the context of influencing personnel to develop an internal dialogue that prompts the question, 'Is it worth the risk?' In the depths behind such a question are personal mores about the sanctity of life, quality of life, financial security as the capacity to sustain earnings over the longer term, and social acceptance within the organisa-tion's culture. Unfortunately, conversations about these values are seldom held and so personnel don't get the opportunity to consider how these fit with their own beliefs or how they align with their employer's culture. These core values can be put into practice through consultative and participative activities led by the supervisor. During onboarding and management of change, there need to be discussions between supervisors and workers that keep team members engaged. Communications shouldn't be limited to technical and social matters. Lack of engagement can cause HR entropic risks to rise due to the risk of cultural degradation. The end-goal should be that the enterprise builds depth of capability, with personnel practising risk management as a life skill (not just doing traditional systems-driven OSH tasks).

Throughout this book, there have been regular discussions highlighting that organisations should properly resource operations to prevent fatalities, serious injuries, and debilitating illnesses. These are the life-changing outcomes of poor operational risk management. That so much effort is put into first aid cases is perplexing, especially given that Heinrich's Triangle has been seriously questioned. There's no correlation between first aid cases and fatalities. As explained by the Entropy Model, high levels of residual risk and/or entropic risk affect the probability of injuries and other serious losses. This points to residual risk management and entropic risk prevention as the critical strategies for preservation of life and protection of workers' quality of life long-term. These critical control verification programmes don't need to be onerous if the workload is shared using a decentralised implementation strategy.

The final part of this chapter explained the Capacity Development Process in Figure 9.4. There were numerous inputs recommended to build organisational capacity starting with an effective recruitment and selection process, and subsequent developmental programmes. These centred on leadership, competencies and the management system with the view to developing a resilient and resourceful organisation. Both enterprise traits are required. The first is to ensure the organisation can adapt to crises and threats whilst the other is a combination of leadership traits that improve job performance. This includes enabling continual improvement and optimising systems of production under controllable, normal risk conditions. Resourcefulness integrates both management and leadership skills that are affective, intellectual and action oriented.

There are numerous expected outputs of organisational capacity, and these influence the enterprise's ability to manage risks and exploit opportunities. The remaining missing piece is to draw lessons learned from successful organisations. In Chapter 10, this concept of capacity building will be explored through the principles of HROs. It's proposed that many organisations and particularly those that are at high risk can learn from HRO's key success factors, as well as from those organisations that considered themselves to be highly reliable but failed miserably as evident by catastrophic events.

REFERENCES

Cambridge Dictionary, 2024, Resourcefulness. English meaning. Cambridge Dictionary.

Covey, S., 1989, *The Seven Habits of Highly Effective People*. New York, USA: Simon and Schuster.

Covey, S.R., 1992, *Principle-Centered Leadership*. New York, USA: Franklin Covey Co., Simon and Schuster.

Department of Energy, Mines, Industry Regulation and Safety (Western Australia), 2024, Officer responsibilities. Department of Energy, Mines, Industry Regulation and Safety (commerce.wa.gov.au).

Foster and Hoult, 2013, Minerals. Free full-text. The safety journey: using a safety maturity model for safety planning and assurance in the UK coal mining industry (mdpi.com).

Fynbo and Jarvinen, 2011, 'The best drivers in the world': drink-driving and risk assessment. *The British Journal of Criminology*. Oxford Academic (oup.com).

Grieve, R., & Van der Stap, T., August 2020, Safety and entropy – a leadership issue. *Professional Safety Journal*, American Society of Safety Professionals, USA.

International Standards Association, 2024, ISO 31000:2009(en), Risk management — principles and guidelines.

Martin-Breen, P., & Anderies, J.M., 2011, Resilience: a literature review. Bellagio-Rockefeller bp.pdf (ids.ac.uk).

Pedaste, M., Maeots, M., Siiman, L.A, de Jong, T., van Riesen, S.A.N., Kamp, E.T., Manoli, C.C., Zacharia, Z.C., & Tsourlidaki, E., 2015, *Phases of Inquiry-Based Learning: Definitions and the Inquiry Cycle*. ScienceDirect.

Sahin, F., Koksal, O., & Ucak, H., 2015, *Measuring the Relationship between Managerial Resourcefulness and Job Performance*. ScienceDirect.

Savvy Calculator, 2024, Experience modification rate (EMR) calculator. Savvy Calculator.

Stephens et al., 2017, *Alcohol Consumption Patterns and Attitudes toward Drink-Drive Behaviours and Road Safety Enforcement Strategies*. ScienceDirect.

Topf, M.D., 2000, Including Leadership in the Safety Process, *Occupational Hazards* March 2000, Penton Media, Inc.

Van der Stap, T., 2008, Overcome the conflict between safety & production with risk management & behavioral safety principles. *RM Insight, American Society of Safety Professionals' Journal*, 8(2). 16–20.

Wikipedia, 2024, Accident triangle. Wikipedia.

Wilde, 1982, The theory of risk homeostasis: implications for safety and health. *Risk Analysis*. Wiley Online Library.

WorkSafe Western Australia, 2021, 211099_GL_WHSOfficers.pdf (www.wa.gov.au).

10 Towards High-Reliability Organisations (HROs)

INDUSTRY GAPS IN RISK MANAGEMENT

As discussed throughout this book, modern organisations are characterised by complexity and uncertainty. They're under constant pressure from stakeholders to demonstrate safety, reliability, productivity and increasing value. They operate in dynamic environments where there's a permanent influx of new and sometimes unanticipated risks due to sociological, technological, organisational, and environmental changes. These external governing and influencing forces were shown in Figure 2.3, the Strategic Alignment Channel. Business entities can become susceptible to unwanted events due to lack of horizon scanning, inadequate risk management, flawed decision-making and deficient processes for managing both known and unknown risks.

In this chapter, much of the discussion will be consolidated with further thought given to what can be learned from successful HROs and those organisations that considered themselves to be high reliability but failed miserably as evident by significant disasters. Several solutions have been proposed in previous chapters including the need for a more mature perspective on the nature of risk. The current understanding and categorisation of risk tends to be static rather than dynamic resulting in insufficient capacity to understand and prepare for uncertainty within a complex and constantly changing environment. Risk is more than a singular concept. It's the potential for loss, harm or negative consequences resulting from uncertain circumstances or actions. There's ambiguity both in the event and the associated consequences, with variability depending on the information and knowledge available. The Society for Risk Analysis argues that it's not realistic to limit the understanding of risk to traditional concepts and has developed a new glossary to allow for different constructs, definitions and measurements of risk (Society of Risk Analysis, 2018).

Uncertainties in risk assessments arise from incomplete information, limited data, assumptions and inherent variability in complex systems. This is confounded by unforeseen events which present a challenge for OSH and operations managers as well as strategists. The occurrence of an unpredictable event can cause highly adverse consequences as found out when COVID-19 arrived and evolved. In Chapter 1, the term 'black swan' was introduced. This refers to unpredictable and unforeseen events that are rare and have a significant impact on society, organisations, or individuals. These were identified as unknown unknowns. Along with known-unknowns, unknown knowns, and known knowns, there's been an attempt to break risks down based on available data.

There are also gaps in risk-based decision-making processes. Common methods include Cost-Benefit Analysis, Risk Acceptance Criteria, Cautionary and Precautionary Principle, and the ALARP principle. These help to justify decisions but don't resolve the dilemmas inherent in organisational decision-making. This was highlighted in the discussions about the Alignment Fallacy and the ALARP Assumption. Because the above practices are well-established and accepted, little has been done to challenge the gaps in society's understanding of risk and enterprise practice. One area of difficulty that was raised was the quandary that engineers face during the project development phase. This affects the levels of residual risk that an operation inherits from the design, construction, and commissioning phases. Such levels may also affect the rates of degradation of technologies and infrastructure over the life of the facility or site. These present ethical challenges for engineers with the outcomes primarily being determined by economic factors, particularly, the ability to attract project investment capital. Competition for investment incentivises shortcuts in design to reduce the total outgoing cost presented to financial markets.

Several major disasters were reviewed in previous chapters. Many of these highlighted the ethical decision-making tensions within engineering design and construction. Flawed decisions were major contributing factors to systems and technology disasters. Such deficiencies may occur in various stages of the project/facilities lifecycle. Some key factors are negligence in engineering design in the name of minimising costs and adhering to a delivery schedule; complex interactions and tensions between engineering and managerial decision-making; and the influence of external organisational and socio-political factors (Van de Poel & Royakkers, 2011).

Major disasters have uncovered evidence of these dilemmas. The system failures resulting in the Challenger and Columbia space shuttle disasters were largely due to the focus on technology, lack of consideration of the social and political system within NASA, and inattentiveness to the consequences of organisational structure and policies on human behaviour and safety (Vaughan, 1996; CAIB, 2003: Chap. 8). The Rogers Commission Report into the Challenger Disaster also identified "flawed decision making" as a contributing cause of the accident. Production and schedule pressures, and violation of internal rules and procedures to launch on time were also significant (Rogers et al., 1986).

The disasters that occurred to the Challenger and Columbia space shuttles are ironic in the sense that NASA was considered a high-reliability organisation. This highlights the vulnerability of enterprises to decay especially under high pressure conditions.

Vaughan (1996) characterises the Challenger launch decision as a case of 'organisational deviance' – values and norms that were seen as unethical outside the organisation, were seen within as normal and legitimate. Faulty decision-making was a common factor in both accidents; however, the decision-makers saw their actions as rational. Enhancing the cultural and organisational environments and appropriate tools were needed to improve decision-making under conditions of uncertainty (Leveson, 2011).

Disasters provide valuable lessons for engineers, regulators, professional societies and educators concerning the ethical responsibilities of their professions. These disciplines have reached the conclusion that there must be a better approach than continuing to learn the hard way. The same applies to operations. Readily available lessons learned abound. In another example, one year after the Hyatt Regency Hotel was completed in Kansas City, Missouri, two walkways suspended over the atrium lobby collapsed in July 1981. More than 200 were injured and 114 people were killed. During the construction stage, the decision was made to attach the set of rods supporting the second-floor walkway to the bottom of the fourth instead of the ceiling. That meant the rods attached to the fourth-floor walkway were supporting twice the weight than the original design intended. The design change was not analysed and approved.

Engineers need to be cognisant of their ethical and professional responsibility (Whitbeck, 2011) but that goes for everyone who makes decisions that affect residual risk levels in the design phase and later when modifications are undertaken on site. Blunders can occur at any time in a facilities' lifecycle if complacency sets in or professional standards are compromised. It's not just systems of production that are subject to degradation, organisational factors can also deteriorate leading to significant vulnerabilities in hard and soft components of the enterprise. These sources of risk have been key discussion points throughout previous chapters.

RISK MANAGEMENT IN HROS

Disasters and failure tend to dominate the headlines, research studies, and indeed, have received a lot of attention in this edition of PSM. There's much to be learned from successful business entities but unfortunately, successes are shared less than failures, and that also tends to be the case in operations.

As explained in Chapter 4 on technology, HROs achieve nearly error-free operations over long time periods and maintain high levels of safety whilst operating in complex, uncertain, high-risk and hazardous environments (Roberts, 1990). They have the capacity to anticipate and prevent undesirable and unexpected events by focusing on the proactive mitigation of risk rather than on responding to errors after these occur. If an unanticipated event occurs, it has the capability to firstly, cope with and contain errors from intensifying and secondly, restoring system functioning (Weick et al., 1999; Weick and Sutcliffe, 2007).

A concept of 'collective mindfulness' is developed giving HROs the capacity to anticipate and contain the unexpected (Weick et al., 1999). There are five principles from effective HROs:

1. Preoccupation with Failure
2. Reluctance to Simplify
3. Sensitivity to Operations
4. Commitment to Resilience
5. Deference to Expertise

The first three principles cover capacity to anticipate and prevent unexpected events whilst the last two are about resilience and containment as a response to threats or crises. The five processes combine to provide a risk-based mindset with processes that are stable and continuously focused to detect the variable patterns of production, activity and response. Early detection enables mitigation of risks of new, unexpected events by coping with unintended consequences (Weick et al., 1999). This aligns with the application of Shannon's Entropy Approach discussed in Chapter 4 about technology. The system is looking for variances not consistencies. The smaller the entropy value the less disorder in the process or systems. In this approach, there doesn't appear to be any judgement about whether the deviance from the expected is a good thing or a bad thing. It just is and draws attention and enquiry.

The fundamental framework of safety should be rethought away from failure at a significant level to also consider day-to-day incremental losses that constitute waste. There may be deviances that can be identified from a mathematical approach to risk and the use of big data or simply, better ways of communicating about the work.

From a safety and efficiency perspective, these wastes can have a cumulative impact.

> Major accidents are often preceded by anomalies – things that are not right, but which have no obvious explanation and, apparently, no undesirable consequence. Too often, anomalies are ignored, until it is too late. One of the vital features of a bad news reporting system is that it can highlight anomalies and ensure that they are responded to in time.

(Hopkins, 2021)

There are two aspects to consider. The first is to stick with what achieves results in terms of risk management. For instance, there are known knowns about what can cause fatalities, for instance, on a construction site, and many of the controls such as 'golden rules' (don't walk under a suspended load) and barriers with procedural controls such as separation of vehicles and pedestrians, are accepted and effective. These should be implemented by every organisation that has these fatal risks as part of its operational risk profile. These controls are legislative and moral obligations.

The second aspect worth considering is greater enquiry into variances that lead to waste. Operationally, waste may include inefficiencies, compromises made because what was needed for the job wasn't available, gaps in communication, gaps in planning, rework and well-intended efforts directed at the wrong outcomes. One way of identifying waste is to ask managers, supervisors and workers what causes frustration in the execution of their work. This will likely inform decision-makers of what's occurring on the rising entropic risk curve, bearing in mind that waste can appear minor but from a systems perspective may be significant.

Waste is an increment of loss.

Waste that's occurring on a rising entropic risk curve can't be recouped.

Time can't be turned back and lost safe production can't be recovered by pushing systems to catch up without introducing more risk.

Waste prevention is a focal point of the Lean approach to continual improvement. Lean can involve complex process mapping to identify deficiencies along the production line; however, in practice, these analytics aren't necessary in under-resourced businesses when waste is often readily identified by supervisors and workers. Internal knowledge is under-exploited in many organisations.

The literature to date that discusses HRO principles highlights their preoccupation with failure. Organisations need to do more than prevent catastrophe. They need to understand operation's prevailing attitude as a starting point. In 'The Simply Irresponsible' enterprise, management will believe failure doesn't exist. 'The Bad' organisation in need of significant transformation will at least have managers who from time to time feel uneasy, and this becomes an avenue that the competent risk professional can explore (without exploitation or fearmongering). HROs are engrossed in actively seeking anomalies, signs of error, small losses and near misses within systems and consider these symptoms of potential system-wide failures.

The culture of HROs encourages workers to report and discuss errors collectively. Near misses aren't treated as evidence of safety rather considered signs of potential failure. Possible vulnerabilities in the system which led to this situation are analysed. In current organisations, wastes are easily overlooked. For instance, a 'broken' procurement system (evident as lack of spares and essential consumables) won't necessarily lead to safety incidents but it will cause inefficiencies, shortcuts, and other deviations from safe, efficient work practices. These are the types of issues that tend to be tolerated rather than reported because there's no glaringly obvious loss. There may, however, be a current of day-to-day frustration amongst operational personnel.

Hopkins (2021) in "A Practical Guide to Becoming a 'High Reliability Organisation'" talks about encouraging bad news reporting and setting up a system to capture and process this information. From operational experience, unfortunately, this may perpetuate characteristics of the OSH and risk-related professions that create a stigma of negativity. A better approach is to report recommended corrective actions, thus aligning reporting systems to the Entropy Model. The Hopkins guide is worth reading and several principles are described:

Principle 1: the reporting technology must be as user friendly as possible
Principle 2: Reports should be routed automatically to particular people
Principle 3: All reports should be individually responded to
Principle 4: Encourage 'helpful' reports
Principle 5: Use local circumstances to steer the system, but not too prescriptively
Principle 6: Encourage courageous reporting
Principle 7: Contractors working on site must be encouraged to participate in the client company's reporting system
Principle 8: There should be no reporting targets
Principle 9: A bad news reporting system depends on top organisational commitment
Principle 10: Governments should guarantee that bad news reporting systems will not increase the risk of prosecution

(Hopkins, 2021)

In terms of practical implementation, many companies already have reporting apps available via mobile phones, tablets and computers, so the development and implementation of such a tool need not be onerous. Essentially, this requires a multidisciplinary approach and user testing. From experience, there can be pushback from some members of the workforce however, most workers are now technologically savvy enough to complete a simple reporting form. For those who say they need more training, a simple response is to ask them how much training they received to do their online banking or shopping on their mobile phone. Resistance doesn't equal incompetence. The tactical risk professional will enlist operational champions to support the introduction of changes to the reporting system, and not surprisingly, suggestions that are listened to and implemented, will create an appetite for fixing day-to-day issues and inefficiencies. This is a good starting point for building the right culture – based on honesty and openness.

In the big picture, HROs are wary of periods of successful operations without any occurrence of errors or breakdowns. This is where OSH performance KPIs such as noticeboards with the number of days since a lost time injury are non-aligned with a risk-based culture. These create a false sense of success that may camouflage the devious tactics used to achieve such outcomes, such as assigning injured workers to administration tasks instead of recording a lost time injury.

The period of good performance was illustrated in the Entropy Model as the optimisation phase (Figure 1.4), which can be followed by degradation when corrective action and maintenance practices aren't sustained (Figure 1.5). As the entropic risk again rises, losses will occur and if left unchecked, will result in an inevitable significant loss. This may be a major safety incident or other catastrophe with severe impact on the organisation's capacity to deliver required results, such as the failure of critical production equipment or a force majeure event. It's therefore important that KPIs are truthful and not create a false sense of security or assurance.

Success can induce over-confidence in the adequacy of current practices and lead workers to drift into complacency and expectation of future repeatable successes. These attitudes and behaviours can increase the likelihood of unexpected events being undetected.

Special attention should be paid to the possibility of potential failures with the assumption that any current success makes future success less probable. While this may sound like paranoia, it should be remembered that these HROs are entities characterised by high risks and potential uncertainties. A sound understanding of the risk profile of such an operation warrants a pre-occupation with failure. In earlier chapters, the risks that facilities inherent from the initial phases of the facilities' lifecycle, specifically, design, construction, and commissioning were discussed. Operation's residual risk profile and tendency towards entropic risk is significantly influenced by project cost decisions. The scenario of minimal project cost and greater risk acceptance is shown in Figure 10.1. There's a high level of residual risk in the operational phase due to deficiencies in HSE in design in earlier phases. The implications are that steps 3 and 4 in the Four-Step Risk Management Strategy become highly important. There needs to be an extreme wariness of residual risks and potential entropic risks to prevent wasteful losses and vulnerability to catastrophe. It's also imperative that

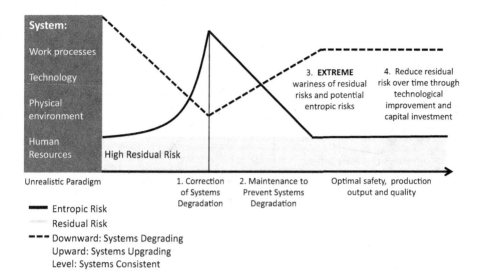

FIGURE 10.1   A minimal project cost decision and higher residual risk for Project A

the operation reduce residual risk over time through technological improvements and capital investments.

> *Some projects are delivered on a lower total outgoing cost for strategic reasons. These include higher likelihood of attracting capital investment. There may be an intention to generate production revenues as quickly as possible to pay down debts and generate cashflow. These decisions aren't wrong per se; however, the company should be aware that operational risks aren't at ALARP.*

Steps 3 and 4 are critical in preventing failures and closing the gaps to where the level of residual risk should have been with a stronger HSE in design basis. The danger is that once production is well underway and revenues are streaming in, there may be limited incentive to make engineering changes that cause disruption to operations, especially, if these have primarily been justified from an OSH perspective. There may be greater motivation if efficiency can also be improved through such investment.

The second scenario shows an optimal project cost decision with a sound HSE basis for design. The residual risk is lower and due to quality componentry and standards throughout the design, construction and commission phases, the tendency towards degradation has a longer lead time. Operations have a lower residual risk level than in the previous scenario. Rather than extreme wariness, facility managers can maintain chronic wariness of the site's risk profile. There's less pressure to invest in capital improvements in the earlier years of the operation compared to the organisation in Figure 10.2.

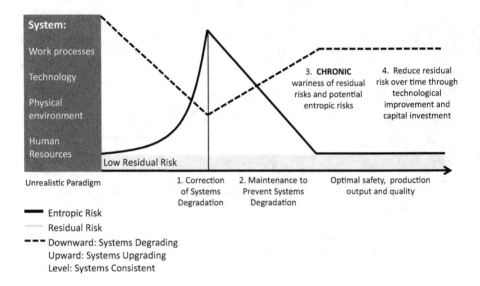

FIGURE 10.2   An optimal project cost decision and lower residual risk for Project B

One of the significant gaps in current risk management practices is that when a site or facility goes into the operational phase, the focus tends to only be on day-to-day risks rather than original deficiencies in design. For instance, during shutdown maintenance of a major hazard facility, decisions will be made to adapt to what is rather than asking how infrastructure and plant should be. As a result, scaffolding may be installed to give maintenance crews access to areas that require upkeep, whereas there should have been ladders, platforms and walkways constructed during the build if the company had been committed to ALARP. An assumption may be made for the purpose of keeping construction costs down that operations will bear these capital expenses.

The organisation intending to adopt HRO principles including preoccupation with failure should literally go back to the drawing board. Much has been written about Prevention through Design (PtD) in practice for the construction industry. There are gaps that Gambatese et al. (2017) identified. Concerningly, these include deficiencies in project designers' focus on the safety of construction workers and minimisation of compliance to legislation such as Construction (Design and Management) Regulations (UK). The commentary should leave construction and operations managers with a sense of unease.

- Much of the focus remains on end-user safety, not construction worker OSH.
- The true benefit comes from working together and communicating; the CDM Regulations were simply the catalyst to enhance project team member communication and integration that ultimately lead to improved safety.
- Changing designer culture in a way that respectfully engages designers recognizes that the burden should not solely be placed on the designer and design culture; placing blame in such a way is not beneficial to the industry and to worker OSH.
- There remains an unclear extent of responsibility for mitigating OSH risks in the design (i.e., how far should a designer go to design out hazards).

- There is an unclear understanding of, and lack of confidence in, the ability for designers to positively impact construction safety as part of their design role.
- It is common to have a misdirected focus on meeting the CDM Regulations instead of implementing PtD.
- The PtD message needs to be expanded beyond the main project team members.

(Gambatese et al., 2017)

If designers have identified these professional gaps, it begs the question, what risks are operations inheriting?

What residual risks and vulnerabilities to entropic risk are current operations managers and their teams exposed to which should be known knowns but may be unknown unknowns due to lack of deep interrogation of the site's risk profile?

Interestingly, the researchers refer to 'changing designer culture' which flags a further segmentation of culture by discipline rather than all stakeholders who determine the total risk profile of an operation having a risk- and opportunity-driven mindset and way of delivering their work.

The preoccupation with failure is likely to lead organisations to adopt a precautionary approach. This means firstly, understanding their residual risk profile and HSE in design deficiencies as described above. Secondly, there should be a constant wariness and need for information about the state of degradation of systems of production. Effectively, the trend should be towards early intervention. This is shown in Figure 10.3 which sees corrective actions brought forward (shown by the solid

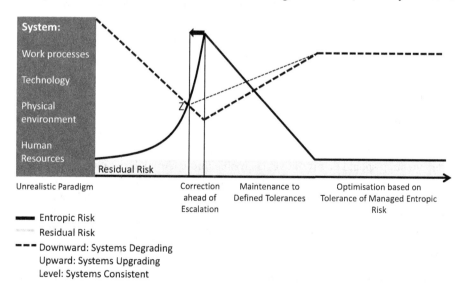

FIGURE 10.3 Decision points due to preoccupation with failure

arrow pointing to the left of the timeline) when the enterprise's risk owners become informed of rising entropic risk within their operations. (The figure builds on figure 4.1 presented in Chapter 4.)

The impact is that systems aren't allowed to degrade as far as otherwise may occur which reduces the accepted level of entropic risk to the level shown at Z. Proactive maintenance practices also commence earlier at Z implying with the fine dashed upward line that upgrades are according to defined tolerances. Optimisation is more sophisticated based on data and adjustments may be made according to agreed parameters.

Many organisations that consider themselves to be HRO may be doing this without the visualisation of risk management shown by the Entropy Model. For less resourced enterprises however, there are some important lessons because in day-to-day decision making, it can be easy to delay fixing safety, maintenance, competence, and information issues which appear to be incremental. It's unlikely that any single manager can confidently assess the cumulative impact of these iterative increases in risk, so it becomes critical that collectively, there's a mindset of unease. This was referred to earlier as 'collective mindfulness'. At the most rudimentary level, this preoccupation with failure (and loss) represents a strong desire to build reliability into the system in terms of delivering safe production to quality requirements.

As mentioned earlier, HROs work to five principles. The second is that these organisations are reluctant to oversimply when interpreting complex situations. This can lead to a false sense of understanding that limits the precautions people take and the various undesired outcomes they can perceive. HROs are aware that to effectively understand the complex environment, they need to build a multifaceted mental model. The discussions about the Entropy Model highlighted the dynamic nature of risk and that current risk registers are at best, a snapshot in time but more often, a list of hazards or work activities. These don't capture the intricacies at the interfaces between systems of production. The use of risk matrices is limited in this regard.

In Chapter 7, the complexity of risk was unpacked by proposing a 'Total Risk Profile' of the organisation as shown by these formulas:

Total Residual Risk Score = $P_{rr} \times T_{rr} \times PE_{rr} \times HR_{rr}$

Total Entropic Risk Score = $P_{er} \times T_{er} \times PE_{er} \times HR_{er}$

Total risk profile = Total residual risks + Total entropic risks

This means that combinations of actual task scenarios can be innumerable; however, attention should be given to catastrophic through to high risks as a priority. As explained previously, multiplication was used (rather than addition) to suggest that a single factor can have a significant impact for an activity. For instance, if the residual risk of technology is high due to lack of safety in design, this can have a major impact on the consequence of an event. Plant built without working at heights addressed during design will expose maintenance workers to a much higher residual risk than if safe, solid access has been installed. From a HR perspective, if the competency level of the workforce is low, this can significantly add to the risk and warrant targeted risk management strategies such as higher levels of supervision. If the HR entropic risk is high due to workload and serious fatigue, this will affect productivity, safety and the quality of the work performed. Mobile and fixed plant that regularly breaks down (i.e. high entropic risk levels), will have a negative impact on production, safety and quality. These trends should be 'red flags' for management's attention. By recognising this complexity, managers may use data to better inform risk mitigation strategies.

The organisation that accepts the complexity caused by the multiple interfaces between systems of production will be reluctant to simplify. This could involve seriously reconsidering the use of a single risk matrix for high-risk processes where these interfaces occur. Amidst these complications, there's an overarching Multidisciplinary Principal Risk and Opportunity Management Strategy.

The objective should be to move all systems of production to higher levels of reliability in terms of delivering safe, efficient, quality work, whilst respecting and being wary of the complexity of the interfaces between these systems.

HROs aim to build capacity and this happens in practical terms by encouraging diversity of perspectives from individuals with different expertise. Better analysis occurs when various aspects of the same event result in a richer and more wide-ranging picture, opening the variety of decision-making options available to address a particular problem. (Weick and Sutcliffe, 2015). By embracing complexity and diversity, HROs are better equipped to navigate and respond effectively to changing demands. A fundamental practice is to apply a multidisciplinary approach to risk management.

From the outset in this edition, the focus has regularly returned to operations which is where the risks 'live'. HROs, as their third principle, are sensitive to operations and maintain a comprehensive understanding of current conditions and mechanics of the system. This is achieved by creating an integrated big picture from continual observations of real-time status (Weick et al., 1999). This allows HROs to comprehend dynamics and pay attention to small deviations or interruptions. These traits provide an interesting reminder about the technological developments associated with the Fourth Industrial Revolution and how big data is assisting enterprises to better understand and manage their risks. Small deviations are also fodder for mathematical applications such as Shannon's Entropy Approach as described earlier. This looks for deviations rather than traditional statistical analysis of what happens most of the time. It's what's different or these subtle variations that should attract management's attention, whilst ensuring that robust critical controls are maintained for known-known, high consequence risks.

HROs are enabled to manage risk effectively by using information systems that collect live data or at least, data with relatively short lag times, to facilitate responsive managerial decision-making. There should be a gnawing curiosity for what is happening operationally – both what is going well to control risk and what isn't going well where iterative losses may be occurring.

The Entropy Model with its Four-Step Risk Management Strategy provides entities with the habits to sustain operational discipline. This in combination with monitoring regimes should enhance sensitivity to operations to prevent significant losses.

In addition, the Hyatt case highlighted how safety in design can be compromised. This should also be a major consideration during operations in relation to modifications to existing infrastructure and/or plant and expansions with interfaces between old and new. The critical questions are, 'How has this changed the residual risk of the operations?' and 'What impact will these changes have on potential entropic risk levels?'.

The final point to consider from an operational perspective is the feedback loop to inform the executive if external or hierarchical internal forces are cascading to the extent that operational discipline is being compromised. In the major disasters described earlier, many root causes were organisational, not just operational. Flawed decision-making was pervasive which indicates the need to have criteria on which to base decisions strategically and operationally. The Reasonableness Test driven by the organisation's Risk Ethos was proposed to address this gap in current business practice. Head office senior management should also seek authentic feedback from operations regarding the practicality and challenges of implementing top-down enterprise strategies, especially if operations are feeling the pressure of change fatigue.

Commitment to resilience is the fourth principle adopted by HROs. This ties in with the Capacity Development Process presented in Chapter 8. Resilience is *"the intrinsic ability of an organisation (system) to maintain or regain a dynamically stable state, which allows it to continue operations after a major mishap and/or in the presence of a continuous stress"* (Hollnagel, 2006). Effective HROs develop capabilities through training and learning that combine knowledge and actions already in their repertoire when dealing with unexpected events. This enhances their range of perception and ability to respond to new threats. Individuals with diverse experience self-organise into ad hoc networks to cope with unexpected events during rapidly developing contingencies (Weick et al., 1999).

There are several obstacles for organisations to achieve HRO status. The categorisation by Aven (2017) of the singular concept of risk remains the basis of ALARP (or SFAIRP) and the fundamental risk process of identification, assessment and control of risk. Unfortunately, operationally after an incident, the tendency in many businesses is to add lower order administration controls to minimise the costs of prevention. The question is whether this is enabling the business to be more resilient or is a hindrance adding increasing reliance on human performance, which is known to be fallible. The other consideration is that additional controls post-incident shouldn't be viewed from only a safety perspective but also explored for improved productivity, quality and/or longer-term cost reduction. The recommended approach, therefore, is to revert to higher order engineering controls that reduce residual risk. This is the intent of step 4 in the Four-Step Risk Management Strategy of the Entropy Model.

Realistically, the reliance on ALARP (which incorporates requisite variety of controls) is unlikely to change from an industry standards perspective. The pragmatic consideration therefore is not to just accept ALARP but to question whether the level of ALARP enables resilience. Logic would suggest that the higher the reliance on human performance for a high-risk scenario, the more tenuous the actual management of operational risks.

'Resourcefulness' should be partnered with 'Resilience' as discussed in Chapter 9. This is the expected overflow or benefit of building organisational capacity.

This was tied in with the Entropy Model and as a result, 'resourcefulness' is intended to lead to better systems review; improved problem-solving and decision-making; effective leadership at all levels; and systems of production raised towards optimal performance levels.

The final principle is that the organisation defers risk-based decisions to expertise. This was explained briefly in Chapter 8 in terms of building organisational capacity. The decentralised delivery of fatal risk critical control monitoring programmes was provided as an operational example, but this principle is more than day-to-day bottom-up strategy. It's a trait that demonstrates a high level of managerial maturity and deep trust between senior management and operations. At the site level, it also means that the operational management team trust the expertise of those subject matter experts in the technical group as well as those directly involved in core business delivery.

In a practical sense, deference to expertise involves adopting a multidisciplinary approach with a common focus on risk management. Hollnagel (2021) referred to the historic 'fragmentation of foci' and the resulting practice of each issue being addressed in isolation without considering the interdependencies. This fragmentation, evident as discipline silos, becomes particularly problematic with the need to manage change.

Concerns about fragmentation should be raised in the context of needed change especially to improve performance. Hollnagel (2021) talks about compensating for degradation due to ever-increasing entropy and the need to recover from disruption.

The Entropy Model promotes deference to expertise according to risk (and opportunity) and collaborative planning and management of complex, inter-related systems. This was described in *PSM1* and this edition as the basis of a strategic, multidisciplinary management system for hazardous industries that ties safety and production together. The principles can apply to all organisations with a sustained concern for the prevention of fatalities, serious injuries and debilitating illnesses, as well as other business losses.

More broadly, organisations seeking to improve their management of risk need to shift away from relying purely on traditional OSH management practices for injury prevention. Within the professional, a more mature perspective is needed that considers systems as complex organisms undergoing dynamic change due to external and internal factors. This doesn't mean however, that traditional methods should be totally disregarded. These may well be enough for low-risk environments with steady state processes. Numerous disasters in high-risk workplaces, however, point to the need for transformation.

The discussions above call for operational risk reviews more often given the dynamic nature of risk, with the active participation of risk owners from various departments, and with a focus on operational challenges and opportunities. The role of the OSH professional in this context is to facilitate consultation to establish a deep familiarity with the total risk profile including interdependencies. The second type of review is for the purpose of management of change to evaluate operational adaptability and readiness. The final type of planned assessment should be to review metrics related to the Four-Step Risk Management Strategies of corrective action, maintenance practices, residual risk management and residual risk reduction. Key questions are: "Has the organisation sustained the agreed standard of operational discipline to ensure risk is being managed effectively? Could the organisation be heading towards failure?" What incremental losses are occurring and can these be rectified and if not, what's the risk?

The final section of this chapter captures some of the key themes and strategies discussed throughout this edition that encourage organisations to reframe their way of working towards a risk-based approach. This means breaking down barriers between silos and defining the common ground. The proposed ten strategies for high reliability consider the risks within systems of production and organisational factors which collectively, make up a Multidisciplinary Principal Risk and Opportunity Strategy. There are likely more elements that can be drawn from the previous chapters and added to this overarching approach. Each enterprise may consider its unique circumstances, capacity and risk/opportunity profile when devising their path to high reliability. In *PSM1*, latter chapters were dedicated to putting together a strategic plan, with various scorecards and setting KPIs for elements within the plan. Strategic planning resources are now readily available via the internet, so this area of general management won't be covered in this edition.

TEN STRATEGIES ON THE PATH TO HIGH RELIABILITY

Throughout this edition, there has been a collective accumulation of strategies and practices to challenge current risk management approaches and propose practical alternatives. This section provides a summary of the key transformation opportunities that have been covered in previous chapters. The issues discussed below include:

1. Transformation of risk-related professions and their work
2. Honesty in organisational identity
3. Understanding and preventing entropic risk
4. Revisiting residual risk and ALARP
5. Understanding the deficiencies of current risk assessment practices
6. Loss investigation and organisational learning
7. Day-to-day operational risk management
8. Risk management as a life skill
9. Capacity through risk leadership and culture
10. The multi-disciplinary approach to risk management

1   TRANSFORMATION OF RISK-RELATED PROFESSIONS AND THEIR WORK

Risk-related professions have come under the spotlight in this edition, with a call to make a transformation from 'safety-based thinking' to 'risk-based thinking'. It's imperative that the OSH discipline in particular addresses conflicts and identity issues by acquiring greater business acumen. Credibility is a high value commodity, especially in hazardous industries and without this, site safety personnel can struggle to influence core business drivers and organisational risks such as production pressure. OSH professionals, having made the transformation themselves, should be influencing managers and supervisors to shift their mindset to risk-based thinking focused on the reliability of systems of production.

Reliability involves safe production delivered to expectation quality standards.

The following is a practical example of how the OSH profession can better serve operations. The Job Safety Analysis (JSA) is a workplace risk assessment tool that is mandated where no safe work procedure exists for the task. There tends to be pushback or covert resistance to using this as it's often seen as a burden and inefficiency, holding personnel back from getting the job done. If however, operational personnel are invited to think about risk, the tool gets workers talking, planning, preparing and risk assessing for a safe, efficient, quality job with the right competencies, tools and communications done ahead of task execution. When these are used should be decided to achieve this outcome. It could, therefore, be called a 'Good Job Plan'. The overarching aim should be to build reliability into the way work gets done to enable continuity of safe production. Naming such tools with a 'safety' adjective undervalues the process from an overarching risk management perspective. This naming practice also differentiates 'safety work' from 'operational work' thus perpetuating silos and perceptions that safety paperwork is a burden hindering the job from being done efficiently.

Organisations can adopt the principles of HROs to achieve business improvement. However, the bureaucratisation of safety and the culture of risk aversion is currently a disservice to the workforce and to progress. Many HSE management systems need an overhaul to ensure that risk mitigations are 'needs' not 'wants'. For this to happen, the OSH Profession should loosen their grip on risk aversion and 'what-if' stories told to justify outdated, redundant approaches to OSH management that border on scare tactics. Conversations need to shift towards what can be done as opposed to what can't be done. Realistically, there are prescriptive parts to OSH legislation which are 'must dos', however, much is left to industry to self-regulate based on the enterprise's duty to workers to the extent that is 'reasonably practicable'. There's great scope for improving safety concurrently with efficiency if this is the approach taken to solving operational risk challenges. As Dekker (2014) points out, OSH management systems should reflect work as done not as imagined. This applies more broadly to operational risk management systems that integrate OSH, environmental management, and quality assurance.

Organisations have been constrained by lack of innovation and critical thinking. Work is dynamic not static. The focus needs to be on risk not simply hazards. Risk-based management systems should be designed to ensure safe, efficient, quality-driven work which concurrently protects operations from chaos and complexity. Such factors should be managed to minimise disruption to operations by the upper echelons of the organisation such as corporate functions. Support services professions should have two main customers – operations and head office.

On site, risk-related professionals should adopt a 360° perspective of their work environment and understand their role to enable operations to work practically whilst also being the conduit that connects operations to the system, including delivering on governance reporting requirements for senior management. If corporate needs collated risk information for business analytics, then the relevant function (OSH, quality, environment) should do this work rather than expecting operational managers to be masters of discipline-driven systems. The frontline should be busy managing operational risks not satisfying the 'system'.

Aspiring HROs should develop lines of accountability between core business and support service functions to clarify who are the risk owners and who are the assurance process owners.

These assurance processes include the internal auditing function and continuous improvement of support service systems and tools. They should also be the focal point for information dissemination, such as changes to legislation, industry news about new incidents from which the business may be able to get 'free' lessons learned, and other trends that improve risk management in the enterprise. The assurance process owners have a key role to play in meeting stakeholder requirements, such as regulators and clients. Finally, a large portion of the work should be in the field providing support and advice.

2   HONESTY IN ORGANISATIONAL IDENTITY

The path to high reliability requires honesty and the demise of organisational silence. Issues that are currently 'undiscussables' should be openly scrutinised. Hopkins (2021) alludes to courageous reporting of bad news. From experience, some of the issues that should be reported as bad news include the rhetoric driven by idealistic philosophy that doesn't sit well with operations. For instance, catchcries such as 'Our people are our Greatest Asset' are as superficial as 'Safety First' when tough decisions must be made about resourcing levels and production targets. Organisations should understand their human capital value to make well informed decisions that impact their HR residual and entropic risks. Likewise, there's no such condition as 'zero' risk. 'Zero Harm' is a fallacy as nil incidents and injuries is unattainable.

Debate about 'Zero', philosophically or statistically, is a distraction from controlling real risks that can have terrible consequences for people, businesses and communities.

Branding of safety has been particularly problematic. 'Safety First' is an ideal that can never consistently survive the operational reality of work. It's tempting to shroud OSH performance expectations in these slogans that appear to uphold high moral values, but the critical piece is whether personnel believe it in the context of their daily experiences. The truth is that 'Safety First' is a lie when workers have no control over the pace of production. An example was given of a food processing company that had a 'Safety First' policy and yet, when a worker was injured on the production line, colleagues failed to assist the person and continued to keep up with the product moving on the conveyor belt. The operation wasn't shutdown. This alarming example high-lighted the depth of cultural issues within the enterprise that stemmed from lack of care and concern for workers openly expressed at senior management levels. It truly takes courage to report this bad news to the Board and it's uncertain if 'The Simply Irresponsible' enterprise can change its ways without wholesale changes at the top.

Many organisations are trying to improve through integrated operational risk and OSH management systems. There can, however, be legacy issues in worker and man-agement relations during the transition phase. A gap is that too often it's assumed by the workforce that senior managers know what the issues are and have failed to act; but management don't always receive truthful reporting from the site level. High level managers simply can't dedicate resources to problems they don't know about or understand. The path to high reliability must involve enquiry into this feedback loop. Further, 'The Work' has always been what matters in operations, not safety branding, symbols, or paradigms. Operations need solutions and these must include both 'hard' tangibles and 'soft' less tangible strategies with genuine intent. Trust comes about more by listening and visible actions than branding, marketing or salespersonship.

It's a necessity that businesses aiming to improve reliability of systems of pro-duction, balance potentially conflicting objectives such as production, quality, OSH performance and cost to ensure short-term viability and longer-term sustainability. The Alignment Fallacy explained some of the significant gaps that erode strategy as it cascades down the hierarchy. This includes the 'Safety First' message and whilst prevention of safety incidents remains paramount, there's a broader concern that should take centre stage.

The key issue that management should be worried about is evidence of losses and whether they have control.

Production downtime, quality deficiencies and safety incidents are all losses con-nected by ineffective risk management. This is overlooked because the definition and approach to risk is singular not dynamic and complex.

Part of the honest leadership discussion should be to acknowledge gaps and opportunities for improvement in industry practice when it comes to the fundamen-tals of risk identification, assessment, control, and verification. A poignant example is the OSH profession's long held reliance on Heinrich's Triangle which predicted a statistical relationship between the number of first aid cases preceding a more serious injury event. For this reason, organisations have expended significant resources on preventing minor injuries. A rethink is required to get to a risk-based approach. The continued occurrence of fatalities, serious injuries and debilitating illnesses indicates that site managers and supervisors need enhanced risk management competencies, better leadership support, coaching and decision-making tools to achieve safe, effi-cient, quality work.

The other significant matter that prospective HROs need to address is regular sense-checks on internal trends. Earlier discussions talked about how leadership and resilience can degrade. Numerous disasters have provided evidence of how organisa-tions can go through cycles of failing, mending, optimising and failing again. This calls for some form of devil's advocacy at senior management levels to ensure that the organisation isn't deteriorating, even though continued performance may suggest otherwise. The challenge may be that people believe what they want to believe. This can be reinforced by groupthink at any hierarchical level. Complacency can set in and then, as the Entropy Model indicates, significant loss can become inevitable.

3 Understanding and Preventing Entropic Risk

PSM1 introduced the concept of entropic risk using the Entropy Model. This risk results from the degradation of systems of production. In organisations that have badly degraded systems, losses tend to affect many functions. From experience of 'The Simply Irresponsible' enterprise, whilst safety-related incidents were high so were personnel issues, environmental incidents, quality deficiencies, re-work, absenteeism and internal conflict. Across all disciplines, the evidence of 'entropic risk' was tangible and expensive. The lesson from this is that a series of serious safety incidents should raise senior management's concerns that there is loss of operational control more broadly across business deliverables not just safety. That's because entropic risk is a collective state of sub-standard systems quality that doesn't just affect safety; it affects everything.

Entropic risk must be part of the language of any organisation aspiring towards higher levels of reliability. Current enterprises may be predisposed to failure simply because entropic risk prevention isn't talked about.

This includes those organisations that considered themselves to be HROs, such as NASA and BP Refinery (Texas,) prior to the catastrophes that befell them.

Due to entropic risk, ALARP should be determined on higher order controls in the hierarchy of controls, being elimination, substitution, and engineering controls. Lower order controls of administration and PPE are vulnerable to variability and sub-optimal performance. Complacency results from the failure to acknowledge that entropic risk is real. The opportunity is to bring entropic risk into the enterprise language and strategy, and to provide managers, supervisors, and workers with practical examples to build understanding of why the quality of their work and decision-making is critical to overall risk management. The focus should always be on 'Good Work' which is safe, efficient and set by clear expectations regarding the quality of the output.

The ALARP Assumption should be embraced by organisations aiming to prevent catastrophes and to reducing losses attributable to systemic decay, complexity, and chaos. Existing operations should view the current approach to ALARP with cynicism. Site management must live with design deficiencies that result from cost minimisation during the design, construction, and commissioning phases of a facility's lifecycle. In addition, as operations mature, there's a tendency after incidents to add more low-cost, lower-order procedural controls that may increase the risk rather than reduce it. Expanding rules based on low value thoroughness as opposed to genuine risk reduction may compromise efficiency and generate new risks such as production pressure.

The entropic risk prevention strategies of corrective action and maintenance practices shown by the Entropy Model, align to the HRO principle of preoccupation with failure. Especially under pressure, an operating environment can slide towards degraded states through sub-standard daily decisions. These include ignoring safety and maintenance issues, leaving the important detail out of risk communications between supervisors and workers, assuming that systems can cope with increasing production demands, and not paying attention to concerns raised by conscientious 'objectors'.

At the most fundamental level, there needs to be an acceptance that entropic risk is real, and no individual can fathom the extent of this within a complex, interdependent system.

The cautious aspiring HRO may start the process of articulating entropic risk and implementing entropic risk prevention strategies in a single work area before rolling it out to the wider site. Inspections, conversations, and internal reports may be used to identify weakly performing areas that can quickly benefit from an upgrade. An effective strategy is to conduct a 'Listening Tour' to engage with the area's managers, supervisors and workers asking them about their greatest sources of frustration when trying to get work done. Fundamentally, this is a question about operational risks, to which there is often also a safety component. An example of this in action was given in Chapter 4 when a site visit was undertaken at a poultry farm, and it was found that rats regularly ate through electrical wiring causing risks to production. The capital investment required to fix the problem was readily attained from senior management because they recognised the problem from a core business perspective. Safety was improved significantly as a result.

Organisations aiming to improve reliability can start with operational risks that affect core business rather than focusing solely on safety, environment, or quality, after prescriptive compliances have been addressed. These functions shouldn't be seen as separate but part of the whole which characterises good performance. Often the real risks that operations face day-to-day that affect efficiency will draw capital investment from senior management's coffers and strategically, the additional costs of HSE in design may be marginal compared to the base cost. The reason this is true in HROs is that reliability and safety go together because both are driven by loss prevention. Some may argue that a machine can be reliable but unsafe or it may be safe but unreliable. From a practical perspective, this doesn't apply in operations. Any serious loss whether an injury or a machine breakdown is a system failure and cause for concern.

In HROs, business improvement can be driven by better operational risk management to encompass efficiency, safety and other benefits. Conversations should be held about how these are inter-related.

The procurement function has a major responsibility to understand operational needs to incorporate safety, efficiency, quality, and reliability when purchasing new plant, equipment and tools. Multidisciplinary input should be part of determining, revising and maintaining procurement standards. This is a key strategy in risk management that affects the residual risk and potential entropic risk of the operations. Such strategy may simply be a refinement for those organisations that have made inroads towards higher levels of reliability.

4 REVISITING RESIDUAL RISK AND ALARP

Extensive discussion has been dedicated to the concept of residual risk and this has entailed a healthy cynicism towards the current definition and practices. There's a universal suite of energies and conditions that can result in workplace fatalities, serious injuries, and debilitating illnesses. Not all will apply in every workplace, but no business is exempt from fatality potential. At this time in society's history, these energy-related risks are known knowns and the required mitigations are public knowledge thanks to legislative requirements, the internet and websites provided by regulators and other key stakeholders. In this regard, there should be no excuses for not having the basic compliance controls in place that keep people safe.

The current gap is that the concept of entropic risk isn't part of industry language and it's assumed that degradation of systems of production are neatly packaged within this residual risk concept. Risk registers will list training, preventative maintenance of technologies, fatigue management and other programmes identified using the Entropy Model as the 'maintenance' phase. Everything is assigned to a pigeon-hole that doesn't consider the interfaces, complexity and potential chaos within these systems and organisational factors. Residual risk as understood by risk owners at each phase of a facility's lifecycle needs to be revisited. The earlier discussion about Prevention through Design raised serious matters about the most fundamental of residual risk mitigations which is front-end engineering and design. This is exacerbated by the cost pressures of getting projects to a financial investment decision. This incentivises lower cost, less HSE in design and cheaper construction methods and as a result, operations may inherit a higher residual risk than currently understood by site manager and OSH personnel. This may not be ALARP except for the potential loophole in the definition of 'reasonably practicable' which accounts for the costs of mitigations and therefore enables lower standards to be justified on shrewd cost-benefit basis.

The probability of an incident relates to the absolute levels of residual and entropic risks at a given point in time, within a dynamic process. For this reason, a written risk assessment becomes redundant as soon as the work commences. Safety then depends on human responses to the changing risk profile. Risk variables must remain within tolerances that people can manage which is tenuous given that human resources are fallible. The real residual risk may therefore be much higher than what management consider it to be. This was the purpose of raising the ALARP Assumption.

HROs with their preoccupation with failure and losses should be wary of ALARP on paper. This warrants a new term: 'ALARP as Documented' (AaD) versus 'ALARP as Material' (AaM).

The term 'material risks' are sometimes used in mature organisations and is a situation when someone or something of value is exposed to danger, harm, or loss. These are often associated with financial costs. Organisations aiming to increase their levels of reliability must be wary about any gaps between their risk assessment and the realistic dynamics that occur within their operations.

The other concern raised throughout the conversation about residual risk is that, if the ALARP Assumption is correct, adding more lower order controls doesn't reduce risk. This means current risk assessment processes are deficient. A new approach that combines residual and entropic risks of an activity, and more broadly, 'The Work', needs to be considered. Decision-makers may not be aware of current risk matrix design features that increase uncertainty. This indicates that the Safety Risk Assessment Matrix is inherently unreliability as a decision support technique. Gaps within other matrices should draw the attention of designers in the first instance.

It may be that current, deficient risk matrices contribute to progressively compounding risk creating unknown unknowns in the built and operating work environment.

An organisation aspiring to be a HRO should go back to the 'drawing board' to better understand their total risk profile. Their objective should be to uncover these unknown knowns. By earlier definition – events aren't known to the analysts who conducted the risk assessment. They're either outside their realm of knowledge or they didn't conduct a thorough risk analysis, but the events are known to others (Aven, 2017). In these circumstances, the risk assessment team should be looking for systems deficiencies that haven't previously been considered rather than trying to determine events. This rationale goes back to the method in Chapter 7 which aims to understand the real risks as they occur in operations. For instance, have there been compromises to procurement standards that mean plant has componentry which is substandard and could malfunction? Perhaps these components were purchased because of cost savings but the result is poor quality and reliability. In the Challenger disaster, it was believed that the failure of the primary and secondary redundant O-ring seals in the right booster caused the catastrophe.

In current operations, are there weaknesses or vulnerabilities that can cause future failure that management aren't aware of?

The organisation adopting HRO principles should want assurance that their residual risks are known knowns. Armed with healthy cynicism, previous risk assessments should be viewed as having procedural flaws, information gaps and inherent misjudgements.

5   Understanding the Deficiencies of Current Risk Assessment Practices

The risk matrix uses likelihood as a key criterion, and this encompasses exposure. This is often understood as a time-related variable but realistically, it's the severity of the risks more so than simply the length of time of contact that matters. Timing becomes important when Reason's Swiss Cheese Model is considered. It's the critical moment when the holes in the system line up leading to a serious injury or incident. In a very high-risk situation, this could occur anytime.

In the discussions, it was proposed that worker exposure is the sum of a complex set of variables associated with systems of production. Collectively, it includes the residual and entropic risks associated with the processes, technologies used and the work environment along with human resources risks. The latter include the residual risks and entropic risks of other workers along with those of the individual/s who may be directly involved in the loss. Current risk assessment practices don't consider this complexity or take a systems approach to breaking risk down into its elements.

In Chapter 7, a different approach was proposed for the mental modelling of risk. One of the primary objectives was to take risk owners and risk-related professionals back to the core strategies that directly affect risk as it exists operationally. The concept of risk modifiers was presented. This is a characteristic of a system of production that either increases or decreases the level of residual or entropic risk. Negative risk modifiers increase the level of risk, whereas positive risk modifiers lower this level. The aim of using risk modifiers is to instil a deep awareness of effective risk management strategies such as HSE in design for residual risks of technologies and the workplace. The risk assessment process should revert to higher order controls rather than imagining more rules or procedures that in most cases, only achieve incremental risk reduction at best.

The eight scorers are designed to be a learning tool to direct risk workshop participants' mindset away from the singular concept of risk and the simplicity of the risk matrix. They should focus on systems rather than behaviours and take a multidisciplinary approach rather than the narrow focus of their discipline (such as safety or quality).

In this edition, from an operational perspective, current OSH risk assessment processes particularly attracted criticism as these are often too much about satisfying the management system than being meaningful to operational personnel. These can involve a group of managers and supervisors in a meeting room going through known hazards or tasks and undertaking a risk assessment with the help of a facilitator. The platform used may be a spreadsheet or database. The outcomes can be far removed from reality especially when there are strong biases towards escalating the importance of one impact over another, for instance, inflating environmental harm of minor spills in non-sensitive areas. From observation, in some cases, it's a means of exaggerating the importance of the work of a particular function. This compromises the process of determining relative risk levels. Instead, the risk assessment should serve two purposes. The first is to create a learning or skill development opportunity for participants and the second is to capture this local knowledge to inform risk-based decision-making.

HROs should be pragmatic about how they undertake risk assessments, so these are meaningful to the people who must own and manage these risks.

The path to high reliability should include a strategy of critiquing and improving on the ways in which risk information is gathered.

6   Loss Investigation and Organisational Learning

When risk isn't the sound basis of professional thinking, cyclical patterns can be repeated like fashions that come and go. Unfortunately, the concept of 'Acts of God' has re-emerged as a root cause of accidents. This needs to be clearly contextualised. The problem is that without accountability, there's no learning from failures or successes, which defies the principles of HRO learning. Calling an incident an 'Act of God' needs to be justifiable. As stated earlier, if it's foreseeable, avoidable, controllable, practical and the organisation's responsibility, it's not an 'Act of God'. Every other undesirable event resulting from the natural environment should be subjected to scrutiny on a case-by-case basis. There's the danger that categorisation of incidents in this way can enable excuses by owners and leaders for failure to act on risk. Don't blame us. Blame Mother Nature.

Senior decision-makers must be alerted to the increasing accountability of enterprise officers due to legislative requirements, environmental and community obligations, and social expectations. This is a matter of fact and has no intended fearmongering tones. Rather than keeping the organisation in the 'survival' phase synonymous with reactive strategies and avoidance of external flack in the pursuit of short-term profit, enterprises intending to remain relevant longer-term must simply accept that compliance is the minimum benchmark. This also requires a different approach to incidents and undesirable events. Decision-makers must accept failings well ahead of any event that attracts the regulators' attention. 'Free' lessons with no or minimal costs to people, the business or the community must provoke a sense of urgency towards reparation. Some have raised concerns that the more an organisation understands its risks and documents these, the more they may be exposed legally if something goes wrong and they've failed to act. HROs are too concerned with prevention to pay undue attention to such 'what if' driven repercussions.

Prevention involves planning for success not just avoidance of failure.
Prevention is also about learning not blaming.

Instead of starting an incident investigation with what the person/s did wrong prior to and at the time of the undesirable event, the question shifts firstly, to analysis of the system and circumstances that led to one or more critical risk decisions. The second consideration is whether those person/s were adequately prepared (by the system) to manage or control the risk state that led to the event. To be technically correct given that 'safety violations' are a component within the organisation's total risk profile, investigators should consider how the organisation failed to protect a competent, conscientious worker; after all, it's the responsibility of the employer to ensure that the worker is adequately prepared to do the work safely.

Aspiring HROs must accept that they control the system that determines
the readiness of personnel to manage risks and to respond to unexpected
contingencies or failures. Competency development is a HRO
accountability.

It's important to accept that most workers (which includes managers) aim to do the right thing for the right reasons. In practice therefore, a 'violation' is not simply intent. It's a matter of whether the behaviour is non-compliant to required practice – procedural (written), the norms of the workgroup (in practice) and the organisational culture (expectations). For a safety violation to have occurred, it must be preceded by individual behavioural risk-taking as the root cause, without extraneous influencing factors that are within the control of the organisation's systems or practices. In HROs, safety violations should be understood within the context of the system.

It's worth reinforcing that it's important to step away from associating risk management with discrimination. The objective is to protect people from harm and enable productive work, not to treat them unfairly or downplay moral and ethical principles. Loss investigations should analyse the interfaces between systems of production and the states of residual and entropic risks of each of these systems. The analysis of person-specific residual and entropic risks should be last as most incidents could have occurred to someone else under a slightly different set of circumstances, such as timing. HRO whilst being a strongly technical entity must balance this systems of production perspective to loss investigation with humanistic values built into those systems, such as procedural fairness.

The focus on human error is a hindrance from exploring fully the underlying variables that lead to losses. These are often attributable to inadequate or degraded systems of production or organisational factors. Few managers have the courage to undertake root-cause analysis into decision-making processes to uncover underlying deficiencies within the management system, culture and/or leadership that could have affected workers' behavioural choices. Poor planning and implementation at the strategic level can result in escalating pressures on operations to achieve targets within constrained timeframes. For this reason, in earlier discussions, organisations were strongly encouraged to support those in influential leadership positions whose performance may be in decline. Learning should encompass having the courage to have conversations about what are currently believed to be 'undiscussable' risks. The overarching objective is to achieve a more objective, systems perspective to understanding loss causation. This is a major shift away from 'unsafe acts' and 'unsafe conditions'.

It's essential for enterprises to mature to a Multidisciplinary Principal Risk and Opportunity Strategy focused on what they know, can uncover, and can control.

Accountability and procedural fairness must be inherent in the way the enterprise is led and managed. This applies at all levels of the hierarchy from company officers to workers on the shopfloor. The implication is that a greater level of conciliation is required with an acceptance of the need to do better whenever failures occur. In the meantime, everyone should be encouraged to not only raise concerns about the safety, quality, and efficiency of the work but also propose solutions. As stated earlier, at the most fundamental level, there needs to be an acceptance that entropic risk is real, and no individual can fathom the extent of this within a complex, interdependent system. It requires a collective effort to learn progressively rather than waiting to learn the hard way.

7   DAY-TO-DAY OPERATIONAL RISK MANAGEMENT

There are numerous synergies gained from having a multidisciplinary approach to organisational development and some of the tried and tested strategies should be revisited. For instance, there are practical benefits for operations from the Lean Production model; however, this should be moderated by avoiding being too pedantic about 'compliance' or 'quality' from an internal auditor's perspective. The truth is that some degree of 'organised chaos' is a natural characteristic of how work gets done. What may look like disorder to the OSH professional who's undertaking a workplace walkthrough may be work as done by operational personnel and it doesn't necessarily mean that conditions are 'unsafe'.

The critical factor is to take a risk-based approach, rather than having false expectations of an ideal world. The internal inspection regime shouldn't be so rigorous to pursue granular detail to the extent that low value thoroughness with greatly diminished return in terms of risk reduction, results. Procedures and checklists in this style invariably deliver negative feedback on non-critical matters. Efforts should focus on real risks. There needs to be a balanced mindset to treat people fairly and not introduce unrealistic expectations that can cause individual or group level psychosocial stress. Regular negative feedback can be demoralising for teams who on the most part are likely trying to do well. More effort should go into providing positive feedback on important risk mitigations and finding solutions for any gaps through consultation with the supervisors and workers of that workplace.

The fundamentals of having a place for everything in its place, once the work has been completed and in readiness for the next phase of production, without creating real risks in the meantime, is a fair and reasonable expectation. Readiness involves restoring systems for the next group of shift workers and providing good communication during a handover process, to achieve teamwork across different interdependent work groups. This should make life easier for supervisors and workers. These are the types of conversations that can be held and lead to improvements. Most people will agree that they prefer to arrive at work to an orderly workplace where they can find what they need to be safe and efficient.

Reliability must extend to how people work together and how they depend on each other to be safe, efficient, and enabled to do a quality job. This applies regardless of the level of the hierarchy at which a person works.

Day-to-day operational risk management should be supported by integrated OSH and operational practice documentation. The 'system' built through consultation and participation drives the desired behaviours and decision-making processes. That's because input builds ownership. At the same time, competencies should be enhanced holistically not by single function. 'Safety' doesn't exist in isolation from other business objectives, especially production so ultimately the aim of life skills training is for personnel to have an internal dialogue that prompts the question, 'Is it worth the risk?' This should be a regular part of daily discussions in HROs at the strategic and workplace levels. Initially, it may be planned through integration into the team meeting agenda so that people become comfortable with these types of conversations, however, eventually, it should occur organically.

On site, operational discipline should be inherent in how systems perform. Discipline doesn't mean punishment in this context, instead it relates to planned, deliberate actions to manage risks. The control of the process comes down to two things. The first is the quality of the systems of production that are being used; secondly, how well the work has been planned, communicated, understood by those involved plus agreed contingencies if something changes during execution that could lead to losses. Operations must be empowered to focus on high-performing core business and critical control management. Supporting functions provide a buffer from influencing forces that are realistically beyond frontline operations' control but can introduce burdensome chaos and complexity.

The path to high reliability involves inclusion of stakeholders who perform the work. This means seamlessness between principal and contractor companies, between the client's and contractors' supervisors, and between employees and outsourced personnel. A risk-based approach involves alignment along the contractual chain, otherwise, risks can be overlooked because of blind spots and assumptions. An example of successful principal-contractor relations was given earlier. Stakeholders made a collaborative effort to deliver an extensive seismic survey on a mine site. The result of the consultative risk assessment was a plan that was agreed by relevant parties and that was worked to. The scope was delivered without incident, in a spirit of goodwill without conflict, on time, on budget and to expected quality standards.

> *The 'perfect job' resulted from 'front-end loading' of risk assessment, planning, and communication with clear roles and responsibilities. Everyone involved had a mindset of what was achievable without being ignorant to the risks.*

The people in the project team perceived each other to be reliable, knowledgeable, and having a common objective of attaining outstanding project delivery. It took a few strong leaders initially to create an environment for success.

> *Aspiring HROs should invest in potential champions to achieve risk leadership competencies across functional areas and vertically through the hierarchy. This investment leads to better 'front-end loading' especially for new projects, scopes of work or ahead of significant changes to the operation.*

8   RISK MANAGEMENT AS A LIFE SKILL

The Entropy Model is designed to give clarity to managers and workers alike about what effective risk management entails. The Four-Step Risk Management Strategy is a life skill that defines workplace practices that are much more meaningful than simply the instruction to comply with the management system. These steps are, in relation to preventing entropic risk, identifying areas of degradation, and taking corrective action. This also involves preventing degradation through well-disciplined operational practices with everyone's input making a difference. What this looks like is that the systems of production that have been designed for safe, efficient, quality work are maintained to agreed standards. If there are recurrent areas of decay, not only are these corrected but also actions taken to implement maintenance practices to raise the quality of these systems to an optimal level. Maintenance means setting, achieving and sustaining systems quality across the four systems of production.

The second part of the strategy is to manage residual risks. At an individual level, high levels of alertness are required when undertaking high risk work or operating within a high-risk environment. At the organisational level, resources are budgeted for on a risk-basis with particular attention to those inherent risks that can cause the greatest consequences if systems fail. Individuals are alert whilst the enterprise maintains chronic wariness and collective mindfulness. The longer-term strategy is to reduce residual risk through capital investment and improvements in industries' knowledge and capacity to manage these risks.

Risk management is a life skill for workers (at work and home) and involves understanding the consequences of failing to manage residual risk and prevent entropic risk. This can be evident at home, for instance, costly residence and vehicle expenses because repairs were delayed. It could be that the person has ignored medical advice to later suffer a more serious health issue. Because risk management is a life skill, there are parallels between risk at the micro and macro levels – individual to organisation to community.

The Entropy Model can illustrate the consequences of leaving entropic risk uncontrolled, with the inevitable result of a serious loss. Along the rising entropic risk curve, management should be concerned about trends pointing to iterative losses that are costing the enterprise directly or indirectly. There's a strong case for adopting the mindfulness principles of HROs. Ahead of any significant loss or crisis, expertise should revert to centres of excellence made up of the subject matter experts with deep understanding of the risks involved.

This is a different approach to training and development than currently exists within industry where each discipline competes for budget. The synergies across safety, environment and quality are driven by risk and opportunity management. Significant saving may be possible from treating these deliverables collectively by building risk leadership capacity.

HROs whilst being preoccupied with failure should also be proactive about reducing waste when resources can be better spent elsewhere on genuine risk reduction and building capacity.

The emphasis of capacity development should be on transferable knowledge and skills that enable a person to adapt and adjust according to the dynamic nature of the risks at work and outside of work. One of the observations of Australian workers is that in the past (50 years ago), they wore shorts, a short-sleeved shirt and anything on their feet when they went to work, even in high-risk environments such as mining. Today, it's not unusual to see people mowing their lawn wearing protective clothing, closed footwear, safety glasses and a hat. What has been learned at work has transferred to the home. A person with these habits isn't going to revert to past standards. This is what organisations should be aiming for – an industry-wide level of risk management competencies.

9   CAPACITY THROUGH LEADERSHIP AND CULTURE

Risk-based professionals feel comfortable with their function-specific plans, procedures, objectives, KPIs, training modules and cultural identity. Often these are developed from their discipline's perspective. Thought should be given, however, to the operations managers, supervisors and workers who are being bombarded with internally driven compliance from multiple disciplines. The growth of self-regulation means that they're expected to believe in, comply with, and deliver results across all these areas as well as core business. Aspiring HROs should be wary of this phenomenon which Hollnagel (2021) referred to as fragmentation.

This also applies to other functions such as human resources management and organisational development. A sense-check is sorely needed to realign various functional areas towards a risk-based, multidisciplinary approach. For instance, there's no such thing as 'Safety Leadership', only 'Leadership'; otherwise, there also must be 'Environmental Leadership, 'Quality Leadership', 'Production Leadership' etc. Rhetoric about what constitutes appropriate leadership has grown. Even the term leadership has morphed into 'Stewardship' in some organisations.

One of the aims of this book has been to look at being a leader from a practical, operational perspective. It was proposed that a continuum of styles, from assertive to collaborative, as it relates to the level of risk and urgency, is a more realistic approach. In this second edition, organisational factors have also been addressed including the need to open the neatly wrapped package of 'undiscussables'. This contains the residual and entropic risks associated with leadership, competencies, management systems and resilience/ resourcefulness. It was proposed that entities have a residual risk because of constraints within these factors which means there's no such thing as perfection only optimisation. Concerningly, these factors can decay setting the business on a path to failure. Both systems of production and organisational factors must be managed to prevent losses but also to achieve continual improvement. Groupthink and stereotyping are threats to enterprise performance, particularly, at the senior levels of the organisation. History has shown this to be the case in many enterprises prior to their catastrophic losses.

The phenomenon of leadership degradation was proposed earlier to be a serious risk. Personnel in influential positions when going through degraded job performance can create a ripple effect that brings the rest of the team down. Just as 'black swans' are unknown unknowns in technologies and the physical environment, 'undiscussables' create blind spots to these issues that can result in loss of capacity to deliver on team objectives. This may be considered too sensitive to discuss; however, the repercussions of inaction warrant a more objective, risk-based mindset balanced with genuine care and concern.

Changes are needed in how leaders are developed. Incumbents should start with a deep breath. Individuals shouldn't pigeonhole themselves into being a certain type of leader or blindly follow a leadership mantra. They should be the best possible version of themselves and develop a deep understand to manage their own risks and opportunities. Risk leadership starts with understanding the operational reality of the organisation. Where possible, seek the truth. During the site visit, the senior management (VIP) party will receive information that is visually and verbally filtered compared to day-to-day operations, leaving them with a sanitised perception of the core business environment. This isn't helpful. They can't fix what they don't know or don't understand. HROs and their leadership team are sensitive to operations, reluctant to simplify and committed to resilience.

Some traditional approaches to leadership development hold true. Transformational leadership requires managers and supervisors to loosen their grip on control without relinquishing it. Their role is to provide a guiding hand and empower their subordinates. Having an agreed, overarching driver of leadership behaviour may be needed to generate synergies across business units and departments. It's worth considering a Risk Ethos, for instance, *"We, [insert organisation name], manage risk to achieve safety, production, and quality concurrently. Every decision we make is for safe production and a quality job. This is our Risk Ethos. This is how we work"*. The rationale behind this strategy is that it allows for situational leadership. For instance, command and control management, whilst not in vogue, isn't wrong. It depends on the risk profile of the work and the capacity of the workforce to manage the risks. More conscious effort is needed to understand and manage capacity constraints. A firm leadership style doesn't mean that the executive team is devoid of care and concern for people. Leadership programmes should be revisited to assess whether these deliver 'Leadership as Imagined' (LAI) or 'Leadership as Needed' (LAN).

Management systems also warrant a rethink. The current dialogue in the OSH profession suggests that these systems are over-developed. From experience, it's more a case that 'needs' haven't been differentiated from 'wants'; the latter resulting from deviating away from a risk-based approach. Unfortunately, this drains resources better spent on fatality, serious injury and debilitating illness prevention. The aspiring HRO would cast a critical eye over their systems to determine if it's a help or a hindrance from an operational perspective. Unfortunately, with over-developed systems, there can be a tendency to add lower order controls after incidents. The enterprise thereby creates its own bureaucracy well beyond what compliance demands – risk management that is 'reasonably practicable'.

Management may be fearful of removing any safety programmes or processes because they committed to them in the first place. At the time, it was probably believed that low value additional procedures would reduce the level of risk further to a new ALARP. This makes it more difficult to turn back the hands of time to a more pragmatic approach. Right-sizing of the management system needs to be done on a justifiable risk basis. Relevant managers and professionals should do a sense-check based on data and to ask operational managers for their input. Some safety commentators have proposed that workers are asked what's the stupidest thing they must comply with in the workplace. Operationally, this is ridiculous. 'Stupid' and 'comply' don't belong in the same sentence. Management would be foolish to opening undermine themselves suggesting to workers that they introduced 'stupid' rules.

Management systems can degrade if not maintained. This can be characterised by increasing levels of disorder and chaos culminating in 'Systems Dementia'. The organisation's memory, cognitive skills, and social structures can fade beyond recognition. There's an additional organisational risk if leaders accept this degradation because, in the short-term, the system continues to provide acceptable performance. Data may still be extracted but there's a danger that no one will questions whether it's 'garbage in, garbage out'. From experience, one of the first opportunities in a new role is to deeply interrogate the management systems using a gaps analysis, well before it becomes familiar. This is the starting point to right-size by collating a strategic improvement plan.

> *HROs welcome a 'Cold Eyes' (Independent Project) Review of their systems to identify excess or waste and make recommendations for down- or right-sizing.*

Organisational factors need to be monitored from a risk and opportunity perspective. A business may have state of the art systems of production today in terms of processes, technologies, physical workplaces, and highly skilled operational personnel, however, if leadership and culture erode at the top, the 'castle' may be destined to collapse. Ultimately, this may come about because of incremental strategic decisions that deviate from the path. At the high level, external influences need to be managed to focus the enterprise on balancing short-term and longer-term objectives.

Much of the discussions in this edition were dedicate to risk-based decision-making processes and the compromises that can occur from project inception to operational handover. There are numerous examples that confirm that enterprises can go from industry leadership to devastating failure. Who (collective not singular) took their eyes off the ball?

10   THE MULTI-DISCIPLINARY APPROACH TO RISK MANAGEMENT

Risk is complex and dynamic, especially in this Fourth Industrial Revolution in which change is the norm and disruption lurks around the corner. Even in steady state operations, the world of work isn't like it was during the Industrial Revolution. The array of risks and opportunities are much broader.

> *Despite technological progress, known knowns remain inadequately managed as evident in harm to people, inefficiencies, and other wastes. Operationally, there's still a strong argument for getting the basics right.*

PSM1 and this edition have provided a more detailed view on the nature of risk with each granular task having eight potential sources – the residual and entropic risks of each of the four systems of production. These collectively, affect the organisation's capacity to achieve multiple objectives including core business deliverables, HSE performance and quality outputs for stakeholders. This spectrum of risk demands a multidisciplinary approach at each touchpoint from the initial concept, front-end engineering and design, through to the built working organisation. Gaps in Prevention through Design were raised to pinpoint where the deficiencies and opportunities for improvement begin, back on the drawing board.

No single discipline can determine the risks inherent in the final build. Once operations commence, the same applies to the management of these risks. A multi-disciplinary approach is fundamental to good management, however, unfortunately, the construction of organisational silos has counteracted this approach in practice. What has happened instead is that some functions such as OSH, because safety is relevant to so many aspects of operations, have grown to the extent that at times, they're thought to be the risk owners and decision-makers. Realistically, this certainly isn't the case. The extensive discussions about the residual and entropic risks of each system of production (Chapters 3 to 6) made this clear. The most significant gap in this regard is that enterprise leaders haven't stepped back far enough to look at risk and opportunity holistically to identify that this is the driver that affects everything, and then ensured implementation top-down and bottom-up strategically in their organisations.

At the site level, the perspective needs to change to well beyond the company gates. In today's business environment, the risks from supply chain and procurement are critical inputs to operational risk management and the organisation's capacity to sustain safe, efficient production. The older style of thinking about the world is to look for knowns, measure it and/or fix it. In the new style, the assumption is not knowing what to look for. What is now more important, is to look for anomalies and unusual trends. In addition, highly complex systems could have unknown unknown risks that lead to total failure. The problem is that traditional approaches to risk assessment won't detect these risks.

There's an urgent need for organisations to rethink their approach to human resources risk management and organisation management more generally. HR risks were discussed extensively as a system of production and within the context of organisational factors. An objective approach was taken without discrimination, objectifying workers, or absence of ethical and moral standards. This is a risk-based approach in action.

Some of the contentious issues in part stemming from the OSH profession, were raised including safety branding, the growth of bureaucratic systems, non-alignment of strategy with operational reality, and the challenges of delivering on safety and production concurrently within management systems that don't address conflicting objectives. All too often, operational managers and frontline supervisors feel trapped between 'a rock and a hard place' when it comes to making decisions about safety versus production. OSH legislation demands that the organisation do what is 'reasonably practicable' to manage operational risks. It's not irrational or arbitrary to have a decision tool called the 'Reasonableness Test' to ensure these leaders in critical operational roles, have a basis on which to determine and justify their day-to-day decisions, concurrently ensuring that they receive the support of the managers they report to.

Sites are where the real risks exist and it's these personnel who are potentially in the firing line when it comes to injuries but also if production isn't achieved. Strategically, organisations can achieve more by considering this perspective. The value of human performance and learning is currently under-appreciated and from a business perspective, under-exploited. This could become an even greater issue in the future with the introduction of new technologies. How will operations personnel raise legitimate concerns about the functionality of AI in the workplace? Will it be trivialised because it's assumed that AI is right? Will senior management be receptive to 'dumb' questions (which may in fact be the best to ask)? Experts are predicting that unknown unknowns will emerge because the traditional approach to 'what if' analysis simply won't work because of the complexity of the interconnectedness of things. A multidisciplinary approach to risk and opportunity management isn't a 'nice to have'. It's a necessity for today's operations.

Back-tracking through this edition of PSM, a key aim was to break risk down so that these multiple inputs can be readily identified according to the risk source within each system of production. It requires more than systems-driven risk ownership. A transformation of mindset is required from safety/ environment/ discipline-based thinking to risk- and opportunity-based thinking. The primary aim of this edition has been to invite the reader to make this transformation.

SUMMARY AND CONCLUDING COMMENTS

This chapter has brought together the contents of previous chapters and invited the reader to make a transformation from silo-based thinking to risk- and opportunity-based thinking. The Entropy Model and other tools within this edition have been combined with HRO principles to devise a Multidisciplinary Principal Risk and Opportunity Strategy to enable this transformation at the organisational level.

These HRO principles were shown to have strengths and weaknesses. It was proposed that it's insufficient to simply have a general preoccupation with failure. The Entropy Model suggests specifically that attention should be paid to incremental losses that constitute waste along the rising entropic risk curve. This calls for proactive habits at the strategic and operational levels of recognising this rise and responding promptly with corrective action but also, implementing maintenance practices for prevention. The model highlights that an organisation for various reasons can go through cycles of failing, mending, optimising and failing again. The phenomenon demands that enterprise leaders be chronically wary of degradation of systems of production and organisational factors. This wariness should always apply to residual risks which are ever present.

Operations were asked to be cynical about their current level of residual risk and vulnerability to entropic risk resulting from decisions made during the earliest phases of their site's lifecycle. False assumptions are readily made that the project delivered after commissioning is at ALARP. This is only justifiable because of the potential loophole within 'reasonably practicable' that considers the cost of risk mitigation. The pressure to obtain financial investment can constrain HSE in design of projects which is difficult to remediate during operations because of disruption and further cost pressures to recoup debts and generate revenue. These residual risks must be managed. The problem is whether these are properly understood and whether facilities have in-built unknown unknowns.

The transformation to risk- and opportunity-based thinking encourages rationalisation of training and development in organisations. Extensive resources are expended on discipline-specific programmes to build 'safety leadership', 'safety culture', 'environmental stewardship' etc. All these functions are risk-based and yet, few if any organisations have consolidated these programmes to take a holistic approach to risk management as an organisational and life skill. There are synergies across functions that are too often prevented by the silos themselves. A multidisciplinary approach would encompass risk competencies, leadership and culture development around the principal strategy.

This edition has focused strongly on enabling operations to concentrate on their key deliverables including critical control identification, implementation and verification. This doesn't just apply to OSH but also continuity of production, environmental management, and quality assurance. Critical control assurance and operational discipline are a must for loss prevention across operational functions. Several opportunities for improvement based on sense-checks were identified. These include 'ALARP as material' instead of 'ALARP as documented' which requires opening new lines of enquiry into operations as built. Management systems should be scrutinised to achieve right-sizing based on risk - 'needs' and not 'wants'.

From an organisation factors perspective, expectations about leadership have resulted in confusion and too many 'hats' for the operations manager and supervisor to switch between. Getting back to basics would entail 'leadership as needed' as opposed to 'leadership as imagined', with influencing personnel enabled to leverage their own personality and style rather than being ascribed some ideal that may be unnatural to that person. These competencies apply along the contractual chain with alignment based on effective risk management to achieve safe, efficient scope execution to the expected output standards. Success is planned for through 'front-end loading' of collaborative risk assessment, planning and ongoing, open communication.

Ten strategies were proposed for moving organisations towards higher levels of reliability. These should be critiqued against the enterprise's needs including those of external stakeholders. To assist this improvement process, the Strategic Alignment Channel may be revisited at any time, to better understand the context of the business and where external governing and influencing forces may affect it. Several tools have been provided to broaden the horizons of the business owner or manager in the SME through to senior managers in large organisations. A risk- and opportunity-based strategy must deliver value in any enterprise, in any context, and to any person who has a stake in safety, production and quality.

The ten strategies are diverse and will apply to various industries and enterprises in different ways. This means that critical thinking should be used when considering the prospects that have been presented. These included: showing honesty in assessing the identity and maturity of the enterprise; understanding risk not as a singular concept but involving residual and entropic risks with complex interfaces; applying caution in relation to current risk management practices; having better ways of managing risk day-to-day that lead to higher levels of participation, consultation and ownership; and customising longer-term strategies to develop capacity and ways of working collaboratively.

From the outset, it's been clear that this edition isn't an academic piece, instead it draws on experience. At times the discussions have swung from strategy to operations with the intention of providing perspective on challenging issues from different angles. The introduction opened with, *"An idealist is essentially risk-loving with respect to the achievement of the ideal"* (Cambridge Dictionary, 2024). The hook of the quote was 'risk-loving' which was necessary to immediate quash any assumption that this book, whilst called 'Productive Safety Management' is a book solely about OSH. It's hoped that the reader has gained much more than a new view on safety. *PSM1* sought primarily to address the conflict between safety and production to achieve improvement using a strategic, multidisciplinary approach to risk management. The title, "Productive Safety Management" reflected the intent at the time. This second edition may have otherwise been better named, "Productive Operational Risk Management".

In this edition, risk and opportunity were presented in equal measure and many of the examples illustrated both aspects of what can be achieved in practice. The original Entropy Model in PSM1 (2003) identified the unrealistic paradigm in which safety, production output and quality are perfect and without risk, and yet branding such as 'Zero Harm' has since risen to prominence and in more recent years, faded away.

In this edition, the Entropy Model was extended to cover organisational factor risks again emphasising the fallacy of perfection. Importantly idealistic expectations of operational managers and supervisors to be perfect leaders able to juggle multiple objectives and wear numerous hats, have been challenged. This is particularly in the context of management systems that haven't addressed inherent conflict between business objectives that such managers are directed to deliver on through mostly lagging KPIs. Solutions such as the 'Reasonableness Test', situational leadership based on risk, and operational risk assessment practices using practical testing and observation, were provided to address this gap.

Almost every operational decision presents a potential dilemma of production versus safety and maintenance issues as described in *PSM1*. Hollnagel (2009) described the Efficiency Thoroughness Trade-off which consolidates this quandary. *PSM1* introduced the Alignment Fallacy as a serious organisation defect especially in relation to 'Safety First' in the context of mixed middle management messaging and operational compromises. The discussion about the Alignment Fallacy was worth revisited. In addition, confusion continues to reign about what operational personnel are responsible for compared to support functions. There's a different between being a Risk Owner and an Assurance Process Owner. If in doubt, ask the question. Some of the 'bad news' may be lack of clarity on accountabilities versus responsibilities.

A key theme throughout this edition has been around what can be done as opposed to what can't be done. Some risk-based professions have embraced the latter mindset because of reluctance to look more deeply into legislative requirements and business opportunities. There are prescriptive components within OSH and environmental management, but there's significant scope for self -regulation. These are avenues for businesses to improve efficiency, safety, and quality concurrently with a refined and adaptive outlook. It's worth having an enquiring mind and asking whether a proposed change is a 'need' or a 'want'. If needed, then it warrants investment which involves getting back to the fundamentals of HSE, efficiency and quality in design. Procedures and wearing PPE rely on people who are fallible with a full range of variability.

Humans are incredible innovators, planners, risk managers and capable of performing miracles if provided with a fertile environment. That's often forgotten in everyday work. There won't necessarily be a major 'win' every week but there can be smaller successes that characterise a job well done, such as a project completed by the team on an agreed understanding of what a 'win' looks and feels like. The same goes for receiving good news rather than being assessed and advised of minor non-conformances that aren't underpinned by realistic risk impact. This book is an invitation to risk-based professionals to be more open to operational trust and human capability. The reason the author has learned so much over the last 20 years since *PSM1* is an unwavering belief that operations have the best solutions to operational challenges and with this belief, turning up to site with an open mind.

Human endeavour should never be forgotten; in fact, it should be at the forefront of management systems. When the 12 boys from the Wild Boars football team were stuck in the Tham Luang Cave in Thailand after being trapped by heavy rains and floods in June 2018, the world watched on. Our collective mindset was that something must be done. It wasn't, it can't be done. This group of children were intent on surviving and were incredibly resourceful and resilient. They were rescued after two agonising weeks because of the incredible dedication of the rescue team, volunteers and the supporting community. The reason this wasn't a miracle was that the outcome was a combination of brilliant risk management, exceptional competence and determination. The story is worth re-reading. It makes us think deeply about our relationship with safety, risk and opportunity.

REFERENCES

Aven, T., 2017, A conceptual foundation for assessing and managing risk, surprises and black swans. In G. Motet, & C. Bieder (Eds.), *The Illusion of Risk Control* (pp. 23–39). Cham, Switzerland: Springer.

CAIB, 2003, *Columbia Accident Investigation Board Report, Volume I*. Washington, D.C.: Columbia Accident Investigation Board.

Dekker, S., 2014, *Safety Differently – Human Factors for a New Era*. Boca Raton: CRC Press.

Gambatese, J., Gibb, A., Bust, P., & Behm, M., 2017, Expanding Prevention through Design (PtD) in Practice: Innovation, Change, and a Path Forward, Joint CIB W099 and TG59 International Safety. Health, and People in Construction Conference Towards Better Safety, Health, Wellbeing, and Life in Construction Cape Town, South Africa, 11–13 June 2017.

Hollnagel, E., 2006, Resilience - the challenge of the unstable. In E. Hollnagel, D.D. Woods, & N. Leveson (Eds.), *Resilience Engineering: Concepts and Precepts*. Burlington, Vt.: Ashgate.

Hollnagel, E., 2021, *Synesis – The Unification of Productivity, Quality, Safety and Reliability*. London: Routledge – Taylor & Francis Group.

Hopkins, A., 2021, A-Practical-Guide-to-becoming-a-High-Reliability-Organisation-Andrew-Hopkins.pdf (aihs.org.au).

Leveson, N., 2011, *Engineering a Safer World: Systems Thinking Applied to Safety*. MA: The MIT Press.

Mol, T., 2003, *Productive Safety Management: A Strategic, Multidisciplinary Management System for Hazardous Industries that Ties Safety and Production Together*. Oxford: Butterworth-Heinemann.

Roberts, K.H., 1990, Some characteristics of one type of high reliability organization. *Organization Science*, 1(2), 160–176. doi:10.1287/orsc.1.2.160

Rogers, W.P., Armstrong, N.A., Acheson, D., Covert, E.E., Feynman, R.P., Hotz, R.B., Kutyna, D.J., Ride, S.K., Rummel, R.W., & Sutter, J.F., 1986, *Report of the Presidential Commission on the Space Shuttle Challenger Accident*. NASA.

Van de Poel, I., & Royakkers, L., 2011, *Ethics, Technology and Engineering: An Introduction.* Oxford: Wiley-Blackwell.

Vaughan, D., 1996, *The Challenger Launch Decision: Risky Technology, Culture, and Deviance at NASA.* Chicago: University of Chicago Press.

Weick, K.E., & Sutcliffe, K.M., 2007, *Managing the Unexpected: Resilient Performance in an Age of Uncertainty.* San Francisco, CA: Jossey-Bass.

Weick, K.E., & Sutcliffe, K.M., 2015, *Managing the Unexpected: Sustained Performance in a Complex World.* NJ: Wiley.

Weick, K.E., Sutcliffe, K., & Obstfeld, D., 1999, Organizing for high reliability: process of collective mindfulness. In B.M. Staw, & R. Sutton (Eds.), *Research in Organizational Behavior* (vol. 1, pp. 81–123). Greenwich, CT: JAI Press.

Whitbeck, C., 2011, *Ethics in Engineering Practice and Research.* 2nd ed. Cambridge: Cambridge University Press.

Index

Printed in the United States
by Baker & Taylor Publisher Services